Continentalizing Canadian Telecommunications
The Politics of Regulatory Reform

In *Continentalizing Canadian Telecommunications* Vanda Rideout examines active political resistance to the radical, neo-liberal transformation of Canadian telecommunications that has been orchestrated by the federal government, big business, and their powerful lobbyists over the last two decades. Rideout focuses on the protection of the public interest, a crucial element neglected by most recent studies, and shows that although alliances have been formed between labour, consumers, and public-interest activists, significant disagreements over issues such as free trade, long-distance and local competition, and a targeted subsidy program for very low-income Canadians have meant that this united front has not been able to counter the forces of the new neo-liberal telecommunications policy regime.

Continentalizing Canadian Telecommunications details the complex relationships between the various corporate and government interests, shows how the changes they brought about have locked Canada's telecommunications system into the orbit of the US system, and discusses the implications this has for Canadians.

VANDA RIDEOUT is an associate professor of sociology at the University of New Brunswick.

Continentalizing Canadian Telecommunications

The Politics of Regulatory Reform

VANDA RIDEOUT

McGill-Queen's University Press
Montreal & Kingston · London · Ithaca

Legal deposit first quarter 2003
Bibliothèque nationale du Québec

Printed in Canada on acid-free paper that is 100% ancient forest free
(100% post-consumer recycled), processed chlorine free.

This book has been published with the help of a grant from the Humanities
and Social Sciences Federation of Canada, using funds provided
by the Social Sciences and Humanities Research Council of Canada.
A publishing grant has also been received from the University
of New Brunswick.

McGill-Queen's University Press acknowledges the support of the
Canada Council for the Arts for our publishing program. We also
acknowledge the financial support of the Government of Canada through
the Book Publishing Industry Development Program (BPIDP) for
our publishing activities.

Statistics Canada information is used with the permission of the Minister
of Industry, as minister responsible for Statistics Canada. Information
on the availability of the wide rage of data from Statistics Canada
can be obtained from Statistics Canada's Regional Offices, its World
Wide Website at www.statcan.ca, and its toll-free access number,
1-800-263-1136.

National Library of Canada Cataloguing in Publication

Rideout, Vanda
    Continentalizing Canadian telecommunications: the politics of regulatory
    reform/Vanda Rideout.
    Includes bibliographical references and index.
    ISBN 0-7735-2425-8
    1. Telecommunication policy – Canada. 1. Title.
    TK5102.3.C3R53 2003   384'.0971   C2002-903179-6

Typeset in Sabon 10/12
by Caractéra Inc., Quebec City

For Andy

who encouraged me
to follow my dream
and supported me each
and every step of the way
Gracie tanto, soul mate.

Contents

Acknowledgements

I would like to acknowledge the assistance of the many people whose help made this book possible. Anonymous reviewers for the Aid to Scholarly Publications Programme and McGill-Queen's University Press provided much-needed suggestions and criticisms. The close and critical readings of the entire manuscript received from Wallace Clement, David Colville, and Vincent Mosco were invaluable. Jim Richardson provided constructive advice on the introduction. Richard Cavanagh, Deborah Harrison, Jennie Hornosty, Barbara Pepperdine, and Sid Shniad read parts of the ever-changing manuscript along the way. Help and patience was also provided by Sharon Cody and Susan Doherty.

Much of the initial stimulation and encouragement that resulted in this work came from my doctoral thesis supervisors, Wallace Clement, Bruce McFarlane, and Vincent Mosco, in the Department of Sociology and the School of Journalism and Communication at Carleton University. The thesis-to-book transformation occurred in an atmosphere of collegiality in the Department of Sociology at the University of New Brunswick, with the help of a small research grant for young scholars. Aurèle Parisien has contributed greatly as my editor and in providing the encouragement to complete this book. Susan Kent Davidson, Joan McGilvray, and Mary Milliken paid careful attention to editing, which helped to improve the manuscript.

Papers based on portions of this material have been presented at the Canadian Communication Association and the Canadian Sociology and Anthropology Association at the Congress of the Social Sciences and Humanities. Questions from these meetings helped to clarify issues about neo-liberal telecommunications policy changes.

The book also benefits from contract research I conducted for the former Department of Communication and Industry Canada. As a result of this research I was able to have many discussions with key policy personnel in both departments and the CRTC. Interviews with representatives of numerous telecommunications businesses, business lobby organizations, telecommunications unions, and non-government organizations such as CAC, FNACQ, NAPO, and PIAC helped to tell the entire story of telecommunications policy reform.

A special thanks goes to my partner and fellow communications scholar Andy Reddick. He provided a calm and caring home life and encouraged me to carry on and finish the work when I was discouraged. Not only did Andy read numerous drafts at both the thesis and manuscript stages; he shared his ideas and concerns about the current state of Canadian telecommunications with me in our long hours of discussion.

All of these people have in one way or another contributed to the final form of the book. The responsibility for this final work, however, rests with me.

Abbreviations

ACTWU	Atlantic Communication & Technical Workers Union
AGT	Alberta Government Telephone (Telus Corporation)
BCE	Bell Canada Enterprises Ltd.
BCNI	Business Council on National Issues
BCOAPO	British Columbia Old Age Pensioners' Organization
BCRL	British Columbia Rail Lightel Ltd.
BC Tel	BC Telephone Company (BC Telephone)
BRC	Board of Railway Commissioners
BTC	Board of Transport Commissioners
CAC	Consumers' Association of Canada
CADPSO	Canadian Association of Data and Professional Software Service Organization
CAT	carrier access tariff
CBA	Canadian Bankers Association
CBC	Canadian Broadcasting Corporation
CBEMA	Canadian Business Equipment Manufacturers Association
CBTA	Canadian Business Telecommunications Alliance
CCC	Communications Competition Coalition
CCF	Co-operative Commonwealth Federation
CCL	Canadian Congress of Labour
CCT	Canadians for Competitive Telecommunications
CCTA	Canadian Cable Television Association
CCWC	Canadian Communication Workers Council
CEP	Communications, Energy and Paperworkers Union of Canada

CFCW   Canadian Federation of Communication Workers
CFIB   Canadian Federation of Independent Business
CICA   Canadian Industrial Communications Assembly
CIPS   Canadian Information Processing Society
CLC   Canadian Labour Congress
CMA   Canadian Manufacturers' Association
CNCP   Canadian National Canadian Pacific Telecommunications (CNCP Telecom)
COTC   Canadian Overseas Telecommunication Corporation (Teleglobe)
CPA   Canadian Petroleum Association
CRC   Communication Research Centre
CRTC   Canadian Radio-television and Telecommunications Commission
CSTA   Coalition pour un Service Téléphonique Affordable (PATS)
CTA   Competitive Telecommunications Association
CTC   Canadian Transport Commission
CTEA   Canadian Telephone Employees' Association
CTG   Canadian Telecommunications Group
CUC   Communications Union of Canada
CUSFTA   Canada-U.S. Free Trade Agreement
CWA   Communications Workers of America
CWC   Communications Workers of Canada
DDD   direct-distance dialled (long distance direct dial)
DOC   Department of Communications
EEO   Electrical Employees' Organization
FAPG   Federated Anti-Poverty Groups of British Columbia
FNACQ   Fédération nationale des associations des consommateurs du Québec
FTU   Federation of Telephone Workers Union
GTE   General Telephone & Electronics Corp.
IBEW   International Brotherhood of Electrical Workers
IDIA   Industrial Disputes Investigation Act
ILO   International Labour Organization
IRDIA   Industrial Relations Disputes Investigation Act
IRPP   Institute for Research for Public Policy
ITAC   Information Technology Association of Canada
IXC   interexchange competitors
LFS   Labour Force Survey
LICO   low-income cut-off
LMS   local measured service
MTS   Manitoba Telephone System
MTS^2 message toll service

MT&T   Maritime Telegraph & Telephone (Nova Scotia)
NAFTA   North American Free Trade Agreement
NAPO   National Anti-Poverty Organization
NB Tel   New Brunswick Telephone (NB Telephone)
NewTel   Newfoundland Telephone Co.
NorTel   Northern Telecom Ltd. (Northern Electric)
OHA   Ontario Hospital Association
PATS   People for Affordable Telephone Service
PCI   price cap index
PIAC   Public Interest Advocacy Centre
PSTN   public switched telephone network
PTTI   Postal Telegraph and Telephone International
RIP   Regulated Industry Program (part of CAC)
SaskTel   Saskatchewan Telephone
STRM   Sindicato de Telefonistos de la Republica Mexicana
TASC   Telephone Answering Service of Canada
TCTS   Trans Canada Telephone System (Telecom Canada –
          Stentor)
TELMEX   Teléfonos de México
TLC   Trades and Labour Congress
TWU   Telecommunications Workers Union
UTW   United Telephone Workers of Canada
VANS   value-added networks
WATS   wide-area telephone service

Continentalizing Canadian Telecommunications

Introduction

Many would argue that the continentalization of Canadian telecommunications began with the Canada-U.S. Free Trade Agreement of 1989. In fact the politics of regulatory reform, and the resistance that developed as part of this, began in 1985 in response to the Macdonald Commission's interim report, which raised concerns about the growing internationalization of business, increasing competition in domestic and world trade, the effects of technology on trade, and the growing protectionism of Canada's largest trading partner, the United States. As a solution, the report argued for a move to "an open market economy" through the implementation of neo-liberal economic policies. To avoid economic isolation, Canada adopted an approach similar to those of the United States, Japan, and the European Community in introducing major telecommunications-service policy and regulatory changes. Similar policy changes occurred in other service sectors such as transportation – airlines, railroads, trucking – and finance: banking and insurance. The major characteristics of these policy changes included liberalization, neo-regulation, privatization, continentalization, and internationalization. What the telecom policy changes have in common with those in these other sectors is that they are the integral service component underlying global capital activity. Consequently, the policy changes associated with each of these sectors have been central to the continental and international expansion of market forces.

Telecommunications services in particular are crucial to the increasingly complex and dispersed global operations of advanced electronic capitalism. What is key for transnational businesses is the free flow of information and data, as well as the transmission of traditional media and multimedia products and services anywhere in the world. Pressures, primarily from large corporate telecommunications users and new telecommunications service providers, have also been reflected in continental and world trade agreements. In both the North America Free Trade Agreement and the World Trade Organization agreement, telecommunications chapters have been included to ensure that there is open trade in this service area. Of particular interest are the enhanced and value-added services covering computerized data, audio, video, and information services. Services are particularly significant for multimedia products because they employ electronic-capital criteria such as computer-stored or restructured processing applications that range from formatting and content to codes. Multimedia products also contain the protocols of transmitted information with telecommunications.

These trade agreements guarantee that private corporate networks will have cross-border access to the public telecommunications networks, including the Internet, so that data, information, and content can flow freely across national borders. Other major concerns for the corporate telecom-service users include cheaper long-distance telephone rates and more choice when it comes to new telecom-delivery services or products. What I mean by telecommunications services is more than a telephone with a connection to one of the major carriers such as Bell Canada, BC Telus, or one of the provincial telephone companies. Rather, a broader understanding of telecommunications services needs to include not only the services and systems of the incumbent carriers but also the new wire-line service providers such as AT&T Canada and Call Net, among others. Telecom services also include the telecom resellers and local-access providers, who may be a telephone or a cable company. Other delivery networks may be provided by wireless, cellular, personal communication services (PCS), or satellite.

This book is about the major changes that have occurred in Canadian telecommunications policy and regulation over the last twenty-five years. It focuses on the extensive political activity and involvement it took to integrate Canadian telecommunications into a North American continental regime.[1] Clement explains that regimes are "structures of power combining capital, labour, the state and popular forces, bound together by hegemonic ideologies and practices ... Regimes are constantly contested formations in which the overall structures of power and the specific powers of the component actors are the subject of struggle" (1994: 96–7).

Regime is used here to mean who is empowered to set the continental telecommunications agenda, and in whose interests. The structures of power for the continental regime combine capital, the state, labour, and popular forces. Together they compose a changing "field of power" (Clement 1994: 96) in which each of these elements is related to each of the others. Struggles developed among domestic and multinational corporate interests, the state, along with federal government departments, the telecom legal and regulatory features, and telecom labour, as well as consumer groups and public-interest advocates. What follows is my account of these struggles as Canada shifted from a Fordist telecom regime to a continental one. I have given particular attention to changes that have arisen and continue to arise from the continental regime by focusing on the implementation of a neo-liberal telecom policy and regulatory model.

I argue that the neo-liberal telecommunications policy model – characterized by liberalization, privatization, and neo-regulation – occurred in the context of a more inclusive process of continentalism. Aspects of this process include a competitive market in telecommunications services; the privatization of federal and provincial publicly owned telecommunications utilities, satellite operations, and international networks; and neo-regulation, which allows for very little public oversight of telecommunications service activity, relying instead on market regulation. These policies bear a resemblance to liberal policies in place at the beginning of the twentieth century, particularly the lack of controls on business and of public accountability. The neo-liberal telecom model also includes the reorganization of existing telecom policy; the centralization of telecom jurisdiction and regulation; and the addition of new telecommunications legislation, while continental and international trade agreements provide additional policy levels by adding trade in services and intellectual property. Concurrently, foreign-investment restrictions have been substantially lifted.

My hope is that this book will bring readers to a deeper critical understanding of the politics of regulatory reform. It aims to do so by focusing on the process of social change and concentrates on how and why policy activists, including the state as a political actor, either supported or resisted the continentalization of Canadian telecommunications. There are three key themes that help to guide this analysis. The first is the corporate forces involved in producing change in the regime, and the second is the key role that the federal government played in setting the agenda for telecom trade and policy reform while at the same time creating hegemonic consent to these reforms. It is the third theme that makes this book different from others, by the attention given to the opposing popular forces and the social impact that the changes have had, and continue to have, on many Canadians.

I draw upon a critical political-economy approach because it provides a more thematic overall account of telecommunications policy reform

by incorporating *all* of the policy players and their struggles over hegemonic consent and resistance. If one explains these changes in terms of
Surtees's metaphor of a "telecom war," this battle was fought not only
by competing businesses but also by corporate telecommunications-
service users, research institutes, federal and provincial governments,
and other warriors, including labour and citizens. This focus on opposition tells the other side of the story, one of tenacious policy resistance.
Why did these groups resist? How did the government deal with this
opposition? The answers to these questions come from an in-depth
investigation of the opposing forces, who include anti-poverty and
consumer organizations, seniors, public-interest advocates, labour
unions, and people who live in the rural and remote areas of the country. Analysis of resistance focuses on the research produced by these
organizations, their policy positions at a number of hearings in the
Senate, the House of Commons, and the Canadian Radio-television and
Telecommunications Commission (CRTC), as well as extensive interviews. This research reveals that telecom inequality is widening, as evidenced by increased local telephone rates for low- and fixed-income
people, and higher rates for basic services for those who live in rural
and remote areas.

BUSINESS SUPPORT FOR NEO-LIBERAL TELECOM POLICY CHANGE

The pro-competition literature identifies two major forces responsible
for telecommunications policy liberalization: technology and business.
When pressed to identify the source of policy change, proponents of
competition tend to focus on the technologies developed by the new
telecom players and how they broaden the range of telecommunications networks, services, and equipment (Noam, 1992; Globerman,
1988). The advent of digital technology and the convergence of computer and telecom technologies have, in this view, contributed to the
pace and diffusion of technologies. Because policy changes are seen as
driven by the technology, the effect of these developments undermines
existing telecom regulatory control (Pool, 1983). This technological-
determinist analysis not only sees technology as self-generative; it fails
to consider the important role that business forces played in changing
telecommunications policy.

Other pro-competition research identifies an extensive list of policy
players in the United States, including the incumbent service provider
AT&T, the new challengers MCI and Sprint, and other political actors
such as academics, government departments, lobbyists, and labour
(Derthick and Quirk, 1985: 35). Horwitz adds the public-interest and

consumer-policy activists (1989: 16). For some, telecom-regulation reform grew from a value-shift that favours the restoration of free enterprise and reliance on competition for settling claims (Derthick and Quirk, 1985: 35). For others the value-shift was specifically about changing public policy so that new telecom providers could compete with the established monopolies such as AT&T (Crandall et al., 1991: 12), or Bell Canada and the British Columbia Telephone Company (Woodrow and Woodside, 1986: 103).

Canadian pro-competition studies of telecommunications policy, by contrast, focus either on the battles between the established telephone service providers and the telecom challengers Unitel and Call Net, or on telecom jurisdiction and regulation that is now outdated (Globerman, 1988; Stanbury, 1986; Woodrow and Woodside, 1986). Some point to powerful corporate forces and large telecom users, but they fail to identify who they are (Janisch, 1986: 338; Schultz, 1988: 5). Janisch is one of the few to raise concern about whether the shift to liberalization would aggravate telecom inequalities, with large corporations ending up as "winners" once long-distance rates were lowered and residential consumers emerging as "losers" because their basic rates would have to increase (1986: 338).

In this book I examine all the corporate forces, such as the traditional telecom service providers, the new telecom challengers, and the large corporate telecom users. I also focus on an area that the pro-competition research has been silent about, the important role that the private and public-funded research institutes and experts played in advancing arguments for competition. My work investigates the large corporate users and provides empirical data on the lobby networks and funding arrangements that often linked these business alliances with the research institutes, adding another level of support for pro-competition ideas. These networks and alliances also had an impact on trade agreements as well as on domestic telecom policy reform. My analysis of the policy players' unequal power positions, however, moves beyond winners and losers, revealing the negative impact that the neo-liberal telecommunications policy shift had on low-income and rural and remote telephone subscribers.

POLITICAL STATE ACTION

Most of the pro-competition research acknowledges the need for governments to change existing telecom policy and regulation. Deregulation, liberalization, and privatization then provide the jolt to reorganize the private and public telecom systems. Phillips views regulatory stagnation as unsuitable for a post-industrial era of integrated telecom and computer technologies (1991: 51). However, most of the research supports only an economic rationale in policy reform. Consequently, policy liberalization and deregulation are presumed to be about changing the rules for a complex competitive market, and it is understood that the new telecommunications and electronic services no longer fit into the old legal and regulatory environment (Bruce, Cunard, and Director, 1986: 3; Stanbury, 1986: 482). Explanations are provided about outdated telecom regulation that took into account natural monopoly, the common-carrier utility principle, and rate regulation (Globerman, 1988: 13). The pro-competition view questions the CRTC regulation that linked cross-subsidies to universal-service goals of accessibility and affordability (Schultz, 1995: 279). Other criticisms of the CRTC include the slow transition to competition (Globerman, Janisch, and Stanbury, 1995: 436). Consequently, the CRTC, as a regulatory body, is considered to be an inappropriate policy vehicle (Woodrow and Woodside, 1986: 119).

The American research that focuses on the politics involved in telecom liberalization and deregulation attributes these policy changes to the formation of new power-brokers: the large corporate telecom users and the new telecommunications providers. In essence, the coalitions that were formed were powerful enough to undermine the traditional telecom powers (Oettinger, 1988: 31). Using economic arguments, this power coalition was successful in garnering support from liberals and public-interest groups to support the divestiture of AT&T (Horwitz, 1989: 6). For others the expansion of corporate telecom users' power can be explained by the volume of telecommunications service purchases and the buying power of the large corporate telecom users. Corporate users were interested in telecommunications systems that would be integrated globally. Furthermore, the powerful telecom users succeeded in weakening the state's capacity to protect domestic policy and regulation (Schultz, 1988: 5, 7). Although the telecom users have been recognized as an important force in the policy process, it is important to note that the Canadian government did not act as a transmission belt on behalf of the multinational players to accommodate global capital expansion.

This book then, provides an extensive investigation of the important role the federal state played in setting the agenda for neo-liberal telecommunications policy changes. I argue that the state was not a neutral player, in the sense that it did not just manage the conflicts of competing telecommunications companies. Nor did the federal government reform telecommunications policy simply in response to the pressures and demands from transnational business. Rather, neo-liberal telecom policy changes occurred over a twenty-five-year period, while acceptance of these changes involved the production of hegemonic consent rather than simplistic determination. I pay careful attention to how business forces, experts, and key federal departments and agencies helped to garner hegemonic consent for those policy changes that included reorganizing domestic policy, while at the same time bringing Canadian telecommunications into a continental orbit. (My particular use of the term "hegemony" finds its roots in its Gramscian [1971] derivative, which emphasizes the *consent, influence*, and *intellectual leadership* used by businesses, experts, academics, and the federal government to implement a neo-liberal telecommunications policy shift.)

I examine how the federal government reorganized and downsized its communications structures, particularly those that dealt with social and cultural issues. This is evidenced in the elimination of the Department of Communication and the integration of Consumer and Corporate Affairs into the restructured department now called Industry Canada. At the same time, departments dealing with economic issues, such as Industry Canada and Foreign Affairs and International Trade, were strengthened. Public regulation and accountability of corporate activity, by contrast, have been dramatically reduced.

Telecommunications policy decision-making, once largely the responsibility of just one federal communications regulator, CRTC, is now actively shared with the federal cabinet, the federal and Supreme courts, the minister of Industry, the Department of Foreign Affairs and International Trade, and the Competition Bureau. Regulatory reform also includes the centralization of telecommunications jurisdiction within the federal government and the introduction of new telecommunications legislation. None the less, the federal government discovered that reorganizing consent for the policy shift was not an easy task. Tensions mounted and opposition developed from provincial governments, labour, consumer organizations, and public-interest advocates.

CANADIAN OPPOSITION

No earlier study has focused on the strong resistance that developed among labour, public-interest advocates, and consumer organizations as a result of neo-liberal telecom policy changes. I analyse this resistance as an indication that the production of hegemonic consent was met in Canada with strong opposition from alternative hegemonic forces. In the United States coalitions were formed between two unlikely groups, conservatives and free-market advocates and liberals and public-interest groups, which together supported the end of the AT&T monopoly (Horwitz, 1989: 16–21). By contrast, Canadian public-interest advocates and consumer organizations fought to maintain affordability and universality. They have since tracked, as predicted, exorbitant increases in the cost of basic telecommunications services as well as increases in telephone drop-off rates for low-income subscribers. Although the CRTC and public-interest advocates are aware of the affordability and drop-off problems, it is still unclear whether anything will be done about it.

The pro-competition advocates would like us to believe that the CRTC imposed a tax on long-distance services through the contribution system to offset the negative impact that cost-based pricing would have on local telecom rates (Schultz, 1995: 280). For the most part, competition has developed in many of the urban centres. Unfortunately, this has not been the case for those people who live in rural and remote areas. Due to a lack of competition the CRTC issued a public notice to discuss the issues associated with service to high-costs areas (Telecom Public Notice CRTC 97–42). As a result of this decision a contribution system (based on an explicit subsidy) has been established for those subscribers who live in the underserviced areas (CRTC 99–16). Even though the CRTC has reformed the lions' share of Canadian telecom policy, eliminating many of the previous subsidies, the commission recognizes, as in the high-cost-service decision, that the economic rationale used to eliminate all telecom subsidies is unrealistic. Under the *Telecommunication Act* the CRTC has a responsibility to residential as well as large-user subscribers.

Labour has also continued to track the growing inequality resulting from telecommunications competition and its effect on downsizing the telecommunications work force. Federal government research has noted that employment in telecommunications services is down 17 per cent from the industry's peak in 1990. Public-interest organizations continue to work towards fairness in terms of access and affordability not only for telecommunications services but also for fairness in terms of access to the Internet. All this leads one to question who has benefited from this imagined competitive telecommunications market.

I also show that the federal government attempted to silence resistance from the forces opposing competition. Silencing resistance at the domestic level was subtle and not easily detectable. It entailed setting a competitive telecommunications policy agenda and rapidly advancing the number of hearings dealing with competitive and regulatory-reform issues. Many of the major public telecommunications decisions made in the 1990s by the CRTC, particularly regarding long-distance and local telephone competition, were written as if there was no opposition from the trade unions, the anti-poverty organizations, seniors, or the consumers' organizations. Moreover, even internal government opposition, such as a dissenting decision by one of the CRTC commissioners, although part of the public record, was still a minority one. A specific federal government strategy was devised to confuse and silence opposition to the neo-liberal telecommunications project, aimed at keeping the opposing groups divided or at keeping them from forming broader alliances with small businesses and the provincial governments.

CONTINENTALIZATION

Increasingly, policies such as the North American Free Trade Agreement (NAFTA) and the Canada-U.S. Free Trade Agreement (CUSFTA) reflect the outcome of internal economic and political elite accommodation in the pursuit of continental and international capital expansion. Pro- free-trade and liberalization research explains that in the 1980s international businesses, the U.S. government, and other developed countries, including Canada, wanted trade liberalization in telecommunications and data services. Free-trade expansion and liberalization was thought to ensure market access to and use of the public telecom carriers not only for the new service providers but for the large corporate telecom users as well (Woodrow, 1989: 123). Others welcomed the inclusion of trade in telecom services in CUSFTA because of the pressure it placed on Canadian policy-makers to harmonize their competition and regulatory policies with the United States (Globerman and Booth, 1989; Janisch, 1989).

I would tend to agree that CUSFTA continentalizes telecommunications, but it does not harmonize Canadian telecom policy and regulation with the United States. If that were the case, then all Canadian telecom service providers would be able to compete, as business discourse has it, "on a level playing field" in all countries. According to the largest Canadian provider, Bell Canada, entering the telecommunications service market in the United States is difficult because of a complicated regulatory environment that includes state public utility commissions, the Federal Communication Commission, the Modification of Final Judgment, and the Department of Justice, as well as other levels of regulation. Thus the different rules and political goals of each country are maintained, but within a neo-liberal competition framework. As such, what we have is integration of neo-liberal telecom competition and regulation, as distinct from harmonization.

Free-trade advocates touted NAFTA as an important vehicle needed to conduct business more efficiently in the competitive international marketplace (Shefrin, 1993: 14). The telecommunications chapter of NAFTA in particular is considered an important step towards integrating what were three different national approaches into a liberalized continental market economy (ibid., 15). Some have even argued that in the move to continental free trade the Canadian government did not go far enough, because the *Telecommunications Act* continued to limit foreign ownership on facility-based carriers to 20 per cent. According to this view, these foreign-investment restrictions would likely lead large corporate subscribers to bypass domestic facilities in their search for efficient new network facilities and management expertise (Globerman, 1995: 241). As this book demonstrates, through holding companies foreign telecom investment can range from 33 per cent up to 47 per cent under World Trade Organization rules. I argue that the major objectives of both NAFTA and CUSFTA entail reducing tariff and non-tariff barriers, relaxing corporate restrictions concerning mergers and takeovers, investments, and capital mobility, and strengthening corporate rights. Viewed from this vantage-point, the move towards freer trade represents both continuity and change (Breton & Jenson, 1991: 206). Both agreements are a continuation of post – Second World War liberalism, but with heightened social conflicts. At the same time, these agreements break with previous national industrial-development strategies (Brodie, 1990) by embracing neo-liberalism and market forces in the drive to integrate Canada further into the North American economy. Canadian telecommunications policy is now located within a continental institutional framework.

Continentalizing Canadian telecommunications has had a number of results. First, another policy level must always be taken into consideration, a level that supersedes federal or provincial policy. Second, in an attempt to build social consent for neo-liberal policies and free trade, the Canadian government created a royal commission to inquire into the country's economic and political problems. The Royal Commission on the Economic Union and Development Prospects for Canada, or the Macdonald Commission, as it is commonly known, moved the country towards an open-market economy and away from a previously mixed economic approach. The Canadian government has often relied on royal commissions to defuse explosive issues, and the Macdonald Commission was no exception. To make sure that the neo-liberal view dominated, however, only the perspective of neo-classical economists shaped the commission (McFarlane, 1992: 293). Although the commission's intent was to garner public approval for continental free trade and liberalization, it accomplished quite the opposite because it inflamed political dissent among labour and other public forces, including community groups, social agencies, churches, farmers, women's and youth organizations, volunteer organizations, the unemployed, and native groups (Drache and Cameron, 1985).

This inability to create public support for neo-liberal policy reform and free trade helps to explain why the state moved in a different direction when it went on to implement both trade agreements. The

Departments of Foreign Affairs and International Trade, Industry, and Communications helped to manage the process that put NAFTA and CUSFTA in place. Continentalization also involved action by Foreign Affairs and International Trade through the establishment of the International Trade Advisory Committee and its subcommittee, the Sectorial Advisory Groups on International Trade (SAGIT). Most of the SAGIT members for Information Technologies and Telecommunications were CEOS and company representatives; only two were communications labour members. Members were privy to free-trade negotiations and could provide input, but all information exchanges were classified and confidential. This secretive aspect of the SAGITs had the effect of silencing labour and preventing public input and accountability to the Canadian people.

The continentalization of Canadian telecommunications is a shift from the "permeable" Fordist telecom regime. Chapters 1 and 2 of this study provide a historical mapping of the Fordist regime. Chapter 1 shows that the telecom policy regulation of a century ago emerged as a response to market failure and growing social problems. The natural-monopoly regulation that continued until the 1980s did not just protect the telephone-service companies; it also required that broad public-interest goals be met, including democratic principles of fair, reasonable rates and universal service.

Chapter 2 examines Canada's permeable Fordist telecommunications regime, which allowed private and provincial ownership of the telephone-service industry and federal and provincial regulation. As the Canadian economy improved in the post – Second World War period, telephone service became more affordable through a flat basic-rate regulation. Although the telephone industry was unionized early on, the unions were weak international (American) and company unions, which benefited the telephone companies. Through the decade following its establishment in 1969, the federal Department of Communication supported a number of universality projects, including expanding telecom service to the North and investing in satellite technology, when the private sector showed no interest. With the creation of a new regulatory body, the Canadian Radio-television and Telecommunication Commission, telecommunications regulation became more inclusive and transparent as hearings opened to the public.

Chapter 3 of this study focuses on the forces that lobbied for telecommunications liberalization. These forces included a number of pro-competition business lobby organizations, such as the Business Council on National Issues. I also analyse how private and publicly funded research institutes contributed to pro-competition ideas and recommendations. The pro-competition recommendations from the business lobby organizations and research institutes paid attention only to the benefits of market interests.

In Chapter 4 I investigate external pressures from the United States and the impact they had on Canadian telecom policy reform. As the first country to liberalize and deregulate its telecommunications service policy, the u.s. became the model that large telecom users in Canada referred to when they demanded similar policy changes and threatened the Canadian government with "bypass" to the American system if their demands were not met. The American government also played an important role by helping with the export of its neo-liberal policy model. At the same time, internal forces within Canada, such as new telecom service providers, challenged the existing telecom regulatory environment and made small advances with liberalization of terminal-attachment and business telecom services.

Chapter 5 analyses the first attempt at policy liberalization, particularly the introduction of long-distance telephone competition in 1985. Although the opposing forces, including telecom labour, consumer groups, and public-interest organizations, helped to prevent competition, their success was short-lived. The federal government took a very pro-active role to advance the neo-liberal telecom policy and regulatory changes. First, an attempt was made to keep the resisting forces divided. Second, the CRTC established a federal-provincial-territorial task force to break the deadlock among the resisting provincial governments. Finally, through a Supreme Court ruling, the federal government succeeded in centralizing telecommunications jurisdiction and regulation.

Chapter 6 examines the strong resistance to the neo-liberal telecom policy shift which consumer organizations, public-interest groups, and telecom labour mounted in the second long-distance telephone-service proceeding. What was different in this second proceeding was the ambiguous position of the Consumers' Association of Canada (CAC). Successive liberalization hearings on local-service pricing options and local competition saw a complete break between the provincial consumer organizations, anti-poverty organizations, seniors groups, and rural groups, and the CAC.

Chapter 7 focuses on the opposing forces' positions regarding the introduction of the Telecommunications Act, which revealed further divisions between telecom labour and the consumer organizations. It also looks at how the Department of Foreign and International Affairs neutralized labour in the free-trade negotiations by creating the Sectoral Advisory Groups on International Trade (SAGITs), including one on Information Technologies and Telecommunications. This SAGIT was responsible for the exclusion of other public-interest and consumer groups from the negotiations altogether. Chapter 7 also explains how

and why NAFTA and CUSFTA add another policy level in continentalizing telecommunications by expanding trade to include telecom services.

The federal government has undertaken the enormous task of changing Canadian telecommunications, and continues to manage the policy
shift. Because I take a critical approach to telecommunications policy
analysis, I am sceptical of claims about the long-term benefits that the
continental telecom regime can bring to all Canadians. Already there
is growing evidence of telecommunications and information-highway
inequalities. More residential telecommunications subscribers have had
to give up basic phone service because they cannot afford it. The move
to cost-based pricing has resulted in the redefinition of high-cost service areas for those Canadians who live in rural and remote areas. To
date this has resulted in the doubling of basic telephone-service rates
and no local competition to speak of. Many telecommunications workers have lost their jobs, while those who are still working do so in
stressful, monitored environments. The neo-liberal telecom policy
model has paved the way for businesses and the federal and provincial
governments to encourage the move to an information/knowledge
economy. This has extended the information and communications
technological gap even further, creating a new gap between those who
have access to the Internet or broadband services and those who do
not, exacerbating class, regional, and local, gender, and age differences.

1 Telecommunications and the First National Policy

> ... Social change is located in the historical interaction of the economic,
> political, cultural and ideological moments of social life, with the
> dynamic rooted in socio-economic conflict.
>
> Wallace Clement and Glen Williams,
> *The New Canadian Political Economy*

With the re-election of the Conservative Party on a platform of economic nationalism, the First National Policy was introduced in 1879. The National Policy became the active instrumental means by which the federal government would intervene in the economy to advance monopoly-capital accumulation; it was the state's way of translating the economic interests of capital into political interests. Consequently, the First National Policy acted as a buffer to protect domestic capital from foreign competition. The state concentrated on creating a national capital system, encouraging national economic development by implementing high tariffs and encouraging western geopolitical expansion and settlement to exploit the new resources in that region. A cornerstone of economic development, railway transportation was considered key to the national project in order to transport western staples such as wheat, timber, and minerals east, to help transport immigrants to settle the west, and to deliver goods from central Canada to a growing western market (Babe, 1990: 42). The first transcontinental transportation systems, the Canadian Pacific Railway and the Canadian Pacific Telegraph, both private companies, were built with massive federal government funding, ostensibly to unite the country politically and economically from east to west (Babe, 1990; Rens, 2001; Winseck, 1998). One of the major contradictions created by the First National Policy, one that contributed to the development of corporate continentalism, was the encouragement of foreign direct investment, particularly foreign investment from the United States, in a number of key sectors such as industry, natural resources, and transportation, as well as the establishment of branch plants in the newly developing telephone sector.

Under the First National Policy a national telephone system developed in what is often referred to as a mixed communication system, made up of private and public telephone companies. At the time, hegemony was based on an alliance among the federal government, indigenous capital dominant in finance, trade, and transportation, medium-sized domestic industrial capital, a petite bourgeoisie in agriculture, and American businesses active in resources, manufacturing, and telecommunications. The new telephone and telegraph technologies were promoted and supported by the federal government because of their importance to capital accumulation and the development of a national economy. But because both technologies were relatively new and still in their development stages, they were included in, and considered part of, the transportation sector. Only ten months separated the formation of the Bell Telephone Company of Canada (1880) from the Canadian Pacific Railway Company (1881). From their inception Bell Canada and the CPR operated parallel enterprises in a highly competitive environment. The many common elements in their early history include their formulation by the same corporate lawyer and their almost identical powers to establish a national communication system (Surtees, 1994: 8). Both businesses were also governed by the same principles, laws, and regulation under the Railway Act until 1992. In these early years the relationships between the telegraph industry, the press, the CPR, and Bell Canada were seldom clear-cut or straightforward.

A strong relationship developed between business and transportation entrepreneurs, arising from a network of investors who were involved in the telegraph and railway industries, the newspaper business, and Canadian merchants as well (Babe, 1990: 37). As Winseck points out, cross-ownership links between the Canadian telegraph industry and the press were also reflected in a shared use of new technology, the harmonization of a legal approach by using the same lawyer, A.B. Aylesworth, and a functional interdependence that produced a hybrid telecom/publishing model (1995: 6). As the most avid users of the telegraph, the press had an insatiable appetite for more news content, with an emphasis on speed and the commodification of time. Concurrently, the telegraph industry relied on the press to generate the business volumes it needed to increase profits. Telegraph operators acted as news correspondents by collecting news, and the telegraph company sold news to the newspapers until 1910[1]. (Babe, 1990: 41; Winseck, 1998: 105). The railway sold exclusive rights to the telegraph industry to construct telegraph lines along its rights-of-way, while telegraph technology provided useful electronic communication that the railway industry used for signalling, control, and safety needs.

The telegraph industry also benefited from state intervention, which included granting a federal charter to the various companies, as well as annual subsidies. Implemented in 1852, the Electric Telegraph Act gave the telegraph companies the right to provide telegraph and telephone service (Rens, 2001: 65). This act by the Canadian government contributed to an ongoing rivalry that would last for more than a hundred years, and continues today between the telephone companies and AT&T Canada.

The railway and telephone companies' relationship was even more complicated. The railway industry had had its eye on the telegraph market for quite some time, and on the telegraph charters permitting the offering of both telegraph and telephone services. According to Rens, Canadian businesses were reluctant to invest in Bell Canada because of its mainly American ownership; they also understood that this new communications business had the potential to weaken their investments in the telegraph industry (Rens, 2001: 66). It was not long before rivalry between the telephone and telegraph companies escalated into fierce competition and outright price wars. Business investors had every reason to be concerned because, by 1909, telephone revenues surpassed those of the telegraph industries (ibid., 193). To complicate matters further, Bell Canada secured several agreements with a number of other railway companies, including the Canadian Pacific Railway Company (CPR), which gave the telephone company exclusive rights to place telephones in CPR-owned railway stations and to construct telephone lines along "railroad rights-of-way" (Babe 1990: 88). Bell's *quid pro quo* to the CPR included free telephone-exchange service among all their offices and stations, as well as free long-distance service to CPR officials. The major benefits for Bell from its contract with CPR were the exclusive rights to construct their telephone lines and place their phones in the CPR stations and along CPR train tracks. According to Babe, it was "this lost provision [that] foreclosed Canadian Pacific Railway Telegraph from entering the telephone business" (ibid.).

It would take almost ninety years before the federal government would permit CNCP Telecommunications (Unitel) to compete with the regional telephone companies to provide telecommunications services. In an attempt to unite the country from east to west, the First National Policy's treatment of the railway, telegraph, press, and telephone industries helped to plant the seeds of continentalization that would germinate later.

THE PRIVATE TELEPHONE COMPANIES

The state, politicians, and the captains of the telephone industry ignored any notion of public oversight as they paved the way for early private telephone company development. From its initial development in the 1870s through to the threat of nationalization in 1905, the newly formed Bell Telephone Company of Canada rapidly grew with the benefit of *laissez-faire* capitalism. First, the company benefited from the federal government's grant of an exclusive charter. The charter allowed Bell to manufacture telephones; to construct, acquire, maintain, and operate telephone systems in Canada; to connect with other telephone companies in the country; to amalgamate with or become a shareholder in any other company; and to construct lines along any and all public rights-of-way (public streets and railways) (Statutes of Canada, May 1882, chap. 9, sec. 4, p. 195). Essentially, the charter permitted Bell to incorporate a private national telephone company. The charter also gave the company the power to pursue telephone development without rate regulation other than applications to Parliament for capital authorization to issue stocks and bonds (Babe, 1990: 68). The charter did, however, stipulate that Bell could not offer telegraph services.

Because foreign investment was encouraged, the National Telephone Company (American Telephone and Telegraph Company [AT&T]) purchased the controlling interest in Bell Canada once the charter was secured.[2] After AT&T obtained Bell's controlling shares, it began to take a more active role in establishing the company's policy direction. This meant that until the 1970s, when Bell Canada bought back all its outstanding shares from AT&T, the company operated for many years as a branch plant of the American telephone company (Smythe, 1981: 141). This was evidenced by a predominantly American board of directors and an American president of Bell Canada, C.F. Sise, who continued to shape Bell in its formative years.[3]

Although the federal government expected that Bell would develop a national telephone company, Bell concentrated its efforts in the central part of the country in order to provide healthy dividends for its shareholders. Armed with charter privileges, Bell began an aggressive campaign of mergers, reorganization, and acquisitions that included the purchase of a number of telephone and telegraph companies, and resulted in dramatic increases in telephone rates so that Bell could recover the costs of its buying spree and power grab.[4] Under Sise's management, however, the company's development was restricted to the lucrative metropolitan areas in the provinces of Ontario and

Quebec. These systems were then integrated through the development of a long-distance network. Part of the company's strategy included aggressive competitive practices and refusing to allow independent telephone companies to interconnect to its system. And it developed its tactical alliances with the telegraph companies to instal its telephone lines on telegraph poles, and with the railways so that it could be the exclusive provider of telephones in train stations. In pursuit of economic security in the lucrative business and residential markets in central Canada, Bell also withdrew from its hinterland operations in the Maritime and the Western regions. The company did, however, maintain controlling interest in the telephone plants in the provinces of Nova Scotia and New Brunswick, but it left their operation to the private provincial companies (Babe, 1990: 74–5).

At the turn of the century Bell's aggressive corporate practices continued to escalate in central Canada. A coalition of municipal governments complained to the federal government that Bell engaged in predatory pricing practices. The independent telephone companies accused Bell of denying them the right to interconnect their phone companies with Bell's long-distance network. The Post Master General of Canada and the municipal governments advanced the idea of nationalizing Bell as a solution to the growing telephone problem.

In 1891, Bell having abandoned any plans to extend its system to the West and the territory of British Columbia, the privately owned Vernon and Nelson Telephone Company Ltd., was granted a charter by the provincial government to operate a telephone company throughout the province (Bernard, 1982). The company subsequently changed its name to the British Columbia Telephone Company (BC Telephone and later BC Tel). Expansion continued as the company bought up many of the independent telephone companies throughout the province. In a fashion similar to Bell Canada, BC Telephone engaged in predatory practices, charging exorbitant and erratic rates as well as denying interconnection to competing independents.

Various municipal governments, including Vancouver and others in the interior of the province, petitioned the provincial government to take over BC Telephone because of these practices. Instead, the provincial government petitioned the federal government for a charter to incorporate a new company, and in 1916 the Western Canada Telephone Company Act was passed (Bernard, 1982: 76). Essentially the act transferred jurisdiction and regulation of BC Telephone from the province to the federal government, which was considered a clever way of getting rid of a provincial problem. This meant that now two private telephone companies, BC Telephone and Bell Canada, were under federal jurisdiction. All other provincial telephone companies, whether they were privately owned, like those in the Maritime provinces, or provincially owned, like those in the prairie provinces, remained under provincial jurisdiction until 1989.

Western Canada Telephone was then amalgamated with the BC Telephone Company Ltd., and the merged company assumed the name BC Telephone Company Ltd. The cost of buying 93 per cent of all the other telephone systems in the province, combined with the expensive outlay for higher-grade equipment and the development of a trans-Canada and international long-distance system, proved too costly for the company. Consequently, in the 1920s BC Telephone was sold to a U.S. company, the National Telegraph and Telephone Corporation (also known as the Anglo-Canadian Telephone Company). Thus, a second American branch plant was established in Canadian telecommunications. In 1955 Anglo-Canadian ownership changed once again when it was sold to the American company General Telephone and Electronics (GTE), under whose ownership it remains. This sale was followed, in the 1990s, by another corporate merger between BC Tel and the privatized Telus (AGT), subsequently renamed BC Telus. Once again, like AT&T's investment in Bell Canada, when BC Telephone came under foreign control, the federal government seemed not to be concerned that another carrier had fallen under foreign control. It would appear that it was acceptable very early on for telecommunications policy to benefit from a continentalist orientation.

THE PUBLIC TELEPHONE COMPANIES

Provincially owned and operated telephone companies developed in the prairie provinces of Manitoba, Saskatchewan, and Alberta because Bell Canada was not interested in exercising its charter to expand a national telephone company to serve the rural areas in the western part of the country.[5] The government of Manitoba sought the advice of telephone expert F. Dagger, who recommended that it gain ownership and control of the telephone companies operating in the province. The expropriation of Bell's operations by the Manitoba government was, however, refused by the federal government. By 1908 the government of Manitoba bought all Bell Canada's holdings in the province, establishing the first public telephone monopoly in Canada, the Manitoba Telephone System (MTS) (Manitoba Telephone Systems, 1990).

Similarly, the Alberta government passed an Act Empowering Municipalities to Establish and Operate Telephone Systems in 1905 (Armstrong and Nelles, 1986). Two years later the province purchased several small independent telephone systems, establishing the second public telephone monopoly in the country, Alberta Government Telephone

(AGT) (Babe, 1990: 110). A large municipal system, Edmonton Telephone, coexisted with AGT until it was purchased by the then privatized company Telus (AGT) in 1994.

The establishment of public telephone companies in Manitoba and Alberta were political decisions that were made by telephone experts and provincial politicians, whereas the decision to establish a public telephone corporation in Saskatchewan was spurred by public concern and input from a number of organizations such as the Dominion Grange, the municipal government of Regina, and the Board of Trade, as well as both provincial political parties. The major complaint from these policy activists was that Bell Canada would not allow independents to interconnect to its system and, what was more important, would not develop rural telephone service (Babe, 1990: 107).

It is important to keep in mind, however, that telephone service throughout most of Canada until the 1950s was beyond the economic reach of most Canadians and considered a luxury (Pike and Mosco, 1986: 18–22). But the low rates and rural expansion projects of the prairie public telephone companies helped to establish social development, social use, and the necessity of the telephone. The establishment of public telephone companies also helped to further the development of social-policy principles of affordability and universality. Moreover, these public telephone systems fostered a public service system that would reach their widely dispersed populations in the small prairie towns and rural areas. None the less, these public corporations were rarely subjected to regulatory oversight, nor did they hold open public hearings. Moreover, the Alberta and Manitoba government telephone companies were primarily established to accommodate domestic and multinational capital operating in both provinces. Just as easily as the two provincial governments made the political decision to establish public telephone companies at the turn of the century, almost ninety years later both decided for political reasons to privatize their telephone companies with little public consultation or accountability. Although the privatization of AGT was met with little resistance, the privatization of MTS was met with public demonstrations in the provincial legislature.

THE FEDERAL AND PROVINCIAL TELEPHONE REGULATORY PATCHWORK

For almost twenty-five years after granting the Bell charter, the federal government did not intervene in Bell Canada's affairs. By the turn of the century Bell's customers had become disgruntled because they were subject to continuous irrational rate increases from city to city, and severe rate variations within different areas of a city or town. People who lived in rural areas, or who were serviced by independent local companies, were denied either service or interconnection to Bell's long-distance network. Independent companies and equipment manufacturers were also locked out by Bell because of its exclusive procurement contracts with its parent company's manufacturer, Western Electric. Many municipal governments under popular pressure for lower telephone rates, and frustrated about the lack of control over the placement of poles and wires on their streets, held plebiscites and voted to establish public ownership of Bell's long-distance network.

Parliament was inundated with petitions from 104 municipalities in 1901 and another 98 in 1902 demanding that the federal government take control of Bell's pricing practices (Commons, *Debates*, 1902: 66). In 1902 the federal government made a feeble attempt to contain this conflict by once again amending the company's Act of Incorporation so that "the rates for telephone service in any municipality may be increased or diminished by Order of the Governor-in-Council upon the application of the Company or of any interested municipality, and thereafter the rates so ordered shall be the rates under this Act until again similarly adjusted by the Governor-in-Council" (Senate, *Debates*, 1902: 131–2). From 1892 until 1907 the federal Cabinet had approved rate and capital increases. C.F. Sise, however, found a loophole that would override the ability of Cabinet to halt the company's rate increases, and opted instead to increase only the rates of new subscribers. The year after the minister of Justice approved the new subscriber rate increases, Bell's request to increase capital moved forward (Rens, 2001: 92–3).

This policy revision was intended to placate the public and relieve pressure on the government. Although it failed to do either, the revision granted Parliament the inclusion of regulatory provisions in the revised charter. The municipalities, and member of Parliament and journalist William Maclean, continued their quest for greater public control of telephone rates. In 1902 Maclean introduced a private member's bill to stop predatory pricing in the telephone industry and initiate public regulation that would change Bell ownership from private to public (Babe, 1990: 92). By now Canadian municipalities and the public had lost patience with the collusive Bell and CPR agreements. The public was also frustrated with the lack of regulation in the industry, which left Bell free to do whatever it liked (Winseck, 1998: 149).

The public-ownership issue continued to gain momentum until 1905, when Parliament received complaints and petitions from 195 municipalities and counties. That same year the federal government decided to investigate Canadian telephone practices through the House of Commons Select Committee Appointed to Inquire into Various Telephone

Systems in Canada and Elsewhere. The committee investigated charges against the Bell Telephone Company that included inadequate service; arbitrary rate discrimination, including high local rates in all cities; high long-distance rates; lack of rural interconnection; the company's use of its entrenched position to stifle effective competition; a lack of public interest; and American control of the company, which disadvantaged Canada (Commons, *Select Committee*, 1905: 5, 7–8). Testimony was heard from fifty witnesses, including representatives of the independent telephone industry, municipal governments, the railway and telegraph industry, the Canadian Manufacturers' Association, the Retail Merchants' Association of Canada, members of Parliament, telephone experts for the committee, and Bell Telephone Company managers, including C.F. Sise, and, from the U.S., AT&T officers and the National Interstate Telephone Association (ibid., Appendix 1: xiii–iv).

Bell and its experts, AT&T, and the Canadian Manufacturers' Association vociferously argued against nationalization. Although strong civic desire and extensive public and political pressures were evident for the establishment of a public telecommunications system, the civic groups and local governments were too fragmented and did not build the alliances and coalitions necessary to mount a sustained alternative hegemonic force to nationalize the telephone company. The federal government, in an ironic move, in fact adopted a continentalist position, placing more credence in Bell and its foreign parent AT&T than in the national players.

Although testimony and letters produced two volumes of written material, the committee's proceedings were never acted upon. The federal government thwarted any attempts at nationalization when the committee chairman, Sir William Mulock, was mysteriously called away. His positions of Post Master General and chairman of the select committee were subsequently filled by A.B. Aylsworth, the lawyer who had represented Bell before the committee. Under Aylsworth the committee ground to a halt, claiming that there was too much evidence and that it was too complex for the committee to come to any conclusion.

Essentially, the Laurier government, responding to the arguments of Bell and its allies, stifled the nationalization issue and postponed the telephone question. When the House of Commons met in 1906, the government presented its solution to the telephone question by introducing a series of amendments to the Railway Act that brought all federally chartered telephone companies under the act's jurisdiction. Section 320 of the act required that a telephone carrier's service rates had to be filed and approved by a regulator, the Railway Board of Governors. Section 321 stated that "all tolls shall be just and reasonable, and shall always, under substantially similar circumstances and conditions ... be charged equally to all persons at the same rate" (Railway Act, R.S.C. 1970, section 321). Section 320 also explained that a carrier could not "unjustly discriminate" or give "undue or unreasonable prejudice or disadvantage" with respect to rates, services, or facilities. By bringing the telephone industry under the transportation mandate through the Railway Act and regulating the company's rates through the Board of Railway Commissioners, the federal government managed the nationalization issue and introduced a watered-down public-service aspect into telephone regulation.

The massive movement to nationalize Bell Canada arising from civil groups, municipal governments, the press, and the opposition party had failed. The transportation industry, the Canadian Manufacturers Association, and AT&T and Bell had succeeded because of their power and interaction with the Canadian government. Directly through the committee hearings and indirectly, Bell lobbied and courted particular members of Parliament to make sure that the company remained in private hands. Some have argued that one of the important outcomes of the Mullock Committee was the establishment of a regulatory body (Babe, 1990: 114; Coulter, 1992: 154; Winseck, 1998: 129–30). As Coulter explains, the introduction of the public-service concept of regulation, which became entrenched in Canadian communications regulation, as it did in other public utilities such as gas and transportation, was one of the concessions made by the federal government and private telephone monopolies to prevent nationalization (1992: 154).

This public-service regulation by the BRC was, however, very weak indeed. For example, Parliament still continued to retain control over the BRC and telephone policy in a number of ways. Bell and BC Telephone came before Parliament to get approval for capital expansion only every ten years. Second, at the first BRC hearing a senior Bell manager, C.F. Sise Jr, was asked about ways to separate long-distance expenses from local ones. Sise Jr arrogantly responded that calculation had never been done and, furthermore, would be impossible to do because both networks were so intertwined (Rens, 2001: 100). When pressed further by the commissioners, Sise Jr. responded that "long distance service was profitable because every call was paid for, and was, at any rate, more profitable than the local businesses in many places" (ibid., 100). Until the BRC hearing Bell had claimed in Parliament that long-distance service was a bottomless pit and that revenues were hypothetical. Clarification of long-distance and basic expense allocations and further evaluation of the company's assets would have to wait for another day when the hearing was cancelled because of the death of the chief commissioner. Subsequently, Bell's rates were accepted, as were the expense allocations. As Rans notes, this was the beginning of many years of the "company's creative accounting practices" (ibid., 100). This creative accounting continued in the future, impacting other areas, including the issue of cross-subsidies between long-distance and basic services, affordability, subscriber drop-off, and basic high costs.

As Winseck explains, "the private interest of the compan[ies] and technological programmes were presented as coequal with the public interest" (1998: 130). Equating the public interest with the provision of more telephones would also result in more profits for the companies. Once public pressure subsided, the federal government initiated the regulation of Bell Canada and BC Telephone in such a *light* way that the companies continued to grow and expand. Despite an increase of 1,700 independent companies, the BRC continued to supported the natural-monopoly principle for federally regulated telephone companies (Babe, 1990: 190; Winseck 1998: 149). Consequently, the public-service aspect of Canadian regulation was part of the reorganization that included producing consent for, and legitimizing the notion of, natural telephone monopoly.

The BRC's commitment to natural-monopoly regulation helped the major telecommunications companies, particularly Bell Canada and BC Tel, to consolidate their control over the telecom industry. The term natural monopoly reflected a broad shift in economic activity away from the free-market approach towards telephone monopolies but combined with government regulation. The Mulock Committee revealed that the new telecom-service industry was unable to sustain a competitive industry structure. Support and justification for natural monopoly reflected a major economic shift. Economies of scale were required to deal with inefficiencies that would occur in duplicating another high-cost network. Natural-monopoly regulation also preferenced systems integrity, which unified an end-to-end telecom network that could be controlled by one company or, in the case of Canada, one in each of the provinces and one in central Canada. In addition, cost-averaging would be necessary to set up a system-wide cross-subsidy program to meet legislative universal-service objectives (Babe, 1990: 137–49; Winseck, 1998: 155–202). Winseck makes the argument that natural-monopoly regulation was a "construct of judicial activism" (1998: 201) that fell out of favour eighty years later once the new telecom providers and telecom users challenged the established monopoly players. Others such as Wilson argue that it was the underlying technical characteristic of the telecommunications industry that led to natural-monopoly failure (1992: 348).

None the less, natural-monopoly regulation led to a second wave of growth and expansion of the telephone industry. By focusing on the location of the key businesses and on the most heavily populated part of the country, Ontario and Quebec, Bell Canada increased its telephone subscribership, and the country benefited from increases in telephone-service diffusion, as Table 1.1 shows. The only anomaly in the growth of telephone service occurred in the Depression years. Between 1931 and 1936 the number of telephones dropped to 1,266,000. Because of economic hard times, the number of business phones dropped by 7 per cent in Bell's territory. At the same time, residential subscriber drop-off was 16 per cent, while the number of rural subscribers decreased by 14 per cent (Pike and Mosco, 1986). This matched the national trends apparent in the other provincial telephone companies, including the public ones in the prairie provinces, despite attempts to maintain cheap subscriber rates. In addition, Bell Canada laid off 8,500 people from 1929 to 1936. However, Winseck notes that this occurred during the same period that Bell introduced its new automated dialing technology (1998: 154). This makes it difficult to ascertain how many lay-offs were due to prevailing economic conditions and how many to technological change.

THE TELEPHONE CONSORTIUM

During the 1920s relations developed between the public and private telephone enterprises, resulting in the formation of the Trans Canada Telephone System (TCTS), an unincorporated non-regulated association. The eight member companies were Alberta Government Telephone, Bell Canada, British Columbia Telephone Co., Manitoba Telephone System, Maritime Telegraph and Telephone Co., New Brunswick Telephone Co., Newfoundland Telephone Co., and Saskatchewan Telecommunications. TCTS's mandate included the construction of an all-Canadian telephone link[6] and another continental link through Bell's use of long-distance lines in the United States (Armstrong and Nelles, 1986: 292). As a co-operative venture, TCTS also managed traffic and divided toll revenues; set technical standards; established the conditions under which telecommunications services would be provided by member companies; established performance indicators for joint marketing functions; set rates; established TCTS as the pivotal entity for negotiating and implementing agreements for international services; and established a system of revenue-sharing through TCTS's clearing house (Armstrong and Nelles, 1986: 292). As a cartel, TCTS assumed enormous power by co-ordinating rates, setting standards, controlling service innovation, discouraging competition, and dictating the pace of change. By the end of the 1930s TCTS was responsible for the integration of the regional private and public telephone systems.

Table 1.1
Number of telephones in Canada 1891–1951

Year	Canada	Bell	Phones per 100 population (%)
1891	24,200	22,700	0.5
1901	63,200	44,200	1.2
1911	354,000	161,200	4.8
1921	919,300	400,300	10.4
1931	1,364,200	774,700	13.0
1936	1,266,200	708,600	11.5
1941	1,562,200	888,300	13.5
1951	3,113,800	1,839,700	21.8

Source: Adapted with permission from Pike and Mosco, 1986: 19.

Consequently all of their systems and assets were valued in roughly the same way. In addition, the consortium members standardized their accounting practices. Over time, even equipment became more compatible, with procurement from Northern Electric (now Nortel), Bell's equipment manufacturer, which became the major supplier to the various provincial companies except for BC Telephone. Such practices resulted in a significant degree of functional integration. Despite the fact that decisions for TCTS activity were based on unanimous agreement, great influence was exercised by Bell Canada, the most powerful member (Instant World, 1971: 70).

TCTS has subsequently undergone two name changes, first to Telecom Canada, and most recently to Stentor. The name changes reflected structural changes, and the consortium has also undergone a significant shift in activities. As well as addressing administrative and economic issues, Telecom Canada/Stentor grew into a powerful political institution with ample resources and expertise to lobby various federal government departments, such as Industry Canada, Heritage Canada, and Foreign Affairs and International Trade. After the introduction of full competition to all areas of telecommunications, including long distance and basic services, the Stentor member companies began competing in each other's regions and markets. This resulted in the break-up of Stentor in the latter part of 1998.

The politics of telephone regulation in Canada under monopoly capitalism produced a mixed system of private and public ownership. What little regulation did exist was fragmented at three different state levels, federal, provincial, and in some cases municipal. In the Maritime provinces of Nova Scotia, Prince Edward Island, and New Brunswick, private provincial monopolies provided telephone service. Provincial regulation was conducted by the Public Utilities Board in each province, which also regulated other services such as electric utilities, bridge tolls, gasoline prices, and northern carrier licences. Provincial regulation did not even occur for a number of years. The public-service aspect of regulation and policy-making was also restricted to decision-making by political and industry elites without any public-interest stakeholder input. And, just as important, when the BRC did regulate the telephone industry, their approval of rate increases and capital expansion helped to develop powerful telephone monopolies.

The federal government's response to radio broadcasting shows just how contradictorily it dealt with communication policy in the 1920s and 1930s. Whereas both federal telecommunications carriers were continentally influenced – when Bell Canada was controlled by AT&T in its formative years, and BC Telephone came under direct foreign (United States) ownership – the radio broadcasting policy that established the Radio Broadcasting Act and the Canadian Broadcasting Corporation (CBC) was specifically established to prevent foreign (read United States) interference or influence. Broadcasting was considered culturally and politically significant; therefore, it was afforded national importance (Raboy, 1990: 45). The resulting social conception of the public broadcasting system developed for a number of years under the supervision and protection of the state, acting in the national interest of all Canadians. This contradictory view of broadcasting and tele-communications as residing in two separate spheres – cultural/political and economic respectively contributed to the continentalist political approach the federal government would take almost sixty years later in its treatment of telecommunications in the Canada-U.S. Free Trade Agreement and the North American Free Trade Agreement. In addition, telecommunications would take on a more continental aspect as the *Telecommunication Act* of 1993 loosened previous limits placed on the foreign ownership and control of a Canadian telecom common carrier, allowing them to rise from 20 to 33 per cent and, as they now stand, to 47 per cent.

As the Canadian public-interest objectives of universality and afford-ability evolved through the twentieth century, and particularly after the Second World War, class and civil conflicts continued to fester.

2 Canada's Permeable Fordist Telecommunications Regime

How the Fordist system was put into place ... depended on myriad individual, corporate, institutional, and state decisions, many of them unwitting political choices or knee-jerk responses to the crisis tendencies of capitalism, particularly as manifest in the great depression of the 1930s.

David Harvey,
The Condition of Postmodernity

By the 1920s the old order of Canadian national economic integration, based on the First National Policy, coexisted with a new, developing order of continental economic integration, which began to intensify in the 1920s and 1930s. It was Harold Innis who observed that industrialization[1] shifted the direction of Canada's trade from an east/west route to a north/south one at this time, for all regions except the prairie provinces (Innis, 1956: 368). Many other significant changes occurred as a massive amount of United States capital was directly invested in key resource sectors. The proliferation of industrial branch plants was evidenced by increases in American investments from one billion to four billion dollars (Granatstein, 1986: 26–8). Combined with the economic and social chaos of the Depression years, the new industrialization was the final straw for the First National Policy. As many medium and small Canadian manufacturers folded, they were either bought out or replaced by subsidiaries of large American conglomerates in a number of core industries such as automobiles and related products, gasoline, machinery, tobacco, electrical products, and aluminum (Clement, 1988: 53). Leading the merger movement, Canadian financial capital helped with this consolidation, concentration, and retrenchment of the Canadian economy, as it would later in the 1980s and 1990s.

Although economic chaos and growing social unrest destroyed the remaining foundations of the First National Policy, it would take almost two decades before the Second National Policy was implemented in 1945. In the reconstruction period after the Second World

War, domestic and foreign capital and the federal government implemented a development strategy that has been referred to as a "permeable Fordist" regime of regulation (Jenson, 1989: 78). A regime of regulation refers to the set of social-policy outcomes that arise from structures of power that combine "capital, labour, the state, and popular forces, bound together by hegemonic ideologies and practices" (Clement, 1994: 96). A Fordist regime of regulation refers to the particular policies, strategies, and compromises that were devised by the state in the post–Second World War period to expand capital accumulation, contain social unrest, and develop a hegemonic environment of consent. Playing an active and central role, the state intervenes in the economy to regulate public-service areas such as banking, air transportation, gas, water, and communications, among others. Ostensibly, labour peace is ensured by the state through compromises that include full-employment policies and guaranteed wage increases. A virtuous economic circle develops as a majority of the population benefit from a rising standard of living and increased disposable income, which is used to buy mass-produced consumer goods, which in turn contributes to economic growth. The implementation of stabilizing social-welfare policies such as pensions for the elderly, unemployment insurance, public education, health insurance, and social assistance,[2] among others, also helps to contain the social contradictions inherent in the capitalist system and at the same time contributes to the hegemonic consent required for the regime. Mahon notes that "Canada's version of Fordism set in motion a process of continental integration. Thus it has indeed been "permeable" to American capital which came to control so much of the resource and manufacturing sector; permeable to the policies of the American state; and permeable to the American labour movement via the role played by the so-called international unions" (1991: 324).

The Canadian Fordist strategy contained two major elements. First, through Keynesian-inspired macroeconomic policy, a Second National Policy was implemented based on increased trade and investment ties between Canada and the United States. Second, Canada's commitment to a more liberalized international trading environment and continental integration was strengthened by exporting its resources and importing capital from the United States. Both federal and provincial governments took on a more active role in the economy, encouraging the exploitation of natural resources such as oil, gas, hydroelectricity, and metals such as silver and gold. Mass consumption played a part in Canadian Fordism, but the accelerator was the inflow of foreign (United States) capital for the extraction and export of natural resources, as well as investment in the consumer-durable industries. Even though the federal government intervened in the economy, it was never totally able to contain unemployment problems because Canada's permeable Fordism siphoned profits out of the economy. Consequently, employment-creation was largely left to the domain of the global strategies of United States multinational corporations (Jenson, 1989: 80). These factors and those in the telecommunications environment helped to strengthen the linkages between the Canadian and American economies. After the capital crisis that developed in the 1970s, economic continentalization would be advanced even more by continental trade agreements such as the Canada-u.s. Free Trade Agreement and the North American Free Trade Agreement.

A PERMEABLE FORDIST TELECOMMUNICATIONS REGIME

Canada's Fordist telecommunications regime was subject to varying degrees of private and state ownership and federal and provincial regulation. The telecommunications consortium's (TCTC/Telecom Canada) hegemonic position was secured by federal and provincial governments through policy defence of "natural" monopolies (Mosco, 1993: 142). Consumer consent grew as universal telephone access increased, as evidenced in the dramatic increases in telephone penetration rates following the Second World War (see Table 2.1).

In only thirty years, concentration in the telephone industry intensified (see Table 2.2). By 1975 there were only 333 telephone companies in Canada, and 18 of these provided 99 per cent of all telephone services. Table 2.2 shows that the number of business and residential telephones doubled every ten years in the post-war period. With the stabilization of telephone service costs and flat-rate local calling, the number of calls made in Canada also doubled every ten years. An increase in disposable income and growing perceptions about national telephone universality were tied, through federal and provincial policy, to the expansion (construction costs) of the telecommunications network. These combined factors also led to an increase of 50,000 full-time telephone workers, mainly operators and installers, by 1975. Benefits accrued not only to the telephone companies but to the provincial and federal governments, to telephone workers, and to consumers.

Telephone affordability, particularly for residential and rural subscribers, was maintained through a program of cross-subsidization, from long distance to local, from urban to rural, from business to residential, and subsidization from Bell and BC Tel to the other regions in the west, the east, and the north. The Fordist era also permitted Canadian telecommunications consumers to benefit from affordable basic phone services through flat local rates that were lower than their

Table 2.1
Canadian telephone penetration
rates 1947–1980

Year	Households with one or more telephones (%)
1947	49.7
1950	60.2
1960	83.3
1970	93.9
1980	97.6

Source: Adapted from Statistics Canada,
Household Facilities and Equipment,
Cat. 64–202; Telephone Statistics,
Cat. 56–203.

Table 2.2
Selected Canadian telephone statistics: trends in the telephone industry 1946–75

Year	1946	1955	1965	1975
Telephone companies	3,056	2,637	2,281	333*
Construction costs ($ million)	–	–	159	534
Number of phones (in thousands)	2,026	4,152	7,445	13,165
Number of telephone calls (in millions)	3,559	6,962	12,440	21,194
Number of full-time employees	33,170	55,673	63,467	82,866

* Eighteen of these companies provide 99 per cent of services.
Sources: Canada, Department of Communication: Annual Report 1977/1978; and adapted from
Statistics Canada, Historical Statistics of Canada, 1990.

counterparts in the United States and most of Europe (Mosco, 1993: 142). But these cross-subsidies also guaranteed a revenue pool to the telecommunications consortium. Telecom Canada/Stentor members only supported the principle of universality as long as their monopoly status was guaranteed by state policies.

The private telephone monopolies in particular benefited from a federal regulatory regime that continued to approve rate increases and capital expenditures. Bell wreaked havoc on the growth of independent telephone companies by continuing to prevent long-distance interconnection. Bell's near monopoly position was strengthened even further between 1950 and 1975, when 375 independent telephone companies in Ontario sold their holdings, reducing the independents to about 5 per cent of the provincial total (Babe, 1995: 190). Moreover, Bell's revenues and assets, as Table 2.3 reveals, doubled every five years. Similarly, BC Tel's power grew as its revenues and assets increased from 1955 to 1980 (see Table 2.3).

Table 2.3
Bell Canada and BC Telephone revenues and assets, selected years 1950–80
($ thousands)

	Year	Total revenue	Total assets
Bell Canada	1955	244,900	945,119
"	1965	592,900	2,519,000
"	1970	936,636	2,865,345
"	1975	1,967,000	5,611,000
"	1980	3,203,100	9,227,800
BC Telephone	1955	32,985	120,330
"	1980	784,759	2,682,324

Sources: Adapted with permission from Bell Canada, *Annual Reports* and *Annual Report
in Form 10-K*; adapted from Statistics Canada, *Telephone Statistics*, Cat. 56-202, 1955–80;
Anglo-Canadian Telephone Company, *Annual Report*, 1980.

TELEPHONE UNIONS

In 1905 the only major opposition to the practices of the telephone industry had come from the municipalities, the independent telephone companies, and some members of Parliament, who together forced the federal government to make limited policy concessions and regulatory changes. Labour opposition, by contrast, was virtually nonexistent. During the period of expansion that began after the Depression years, Bell expanded its work force, providing steady employment with reasonably good wages, pension benefits, and virtually no lay offs (Winseck, 1998: 162–4). These employment practices and Bell's anti-union strategy made it very difficult for any union to organize the work force. At the same time the municipalities were waging war on the telephone companies, the international trade unions were in the process of organizing some of the telephone workers in Canada. Significant dissent and resistance from labour did not develop until a mature telecommunications Fordist regime had developed. For almost seventy years the weak position of labour contributed to the hegemonic consent that was forged among the state, telecommunications users, and suppliers, resulting in the telecommunications monopoly. But the private telephone companies would also create the conditions that would lead to further unionization and labour radicalism. The telephone unions. then, are the first socially progressive movement to initiate any effective resistance to the telecommunication industry.

From the turn of the century until the 1970s the international unions struggled to organize telephone workers and to bargain for improvements to wages, conditions of work (hours), and job classifications (Bernard, 1982; Winseck, 1993). This trade-union activity is indicative of what Glenday calls "business unionism" (1994: 27), which is characterized by concern over wages, with particular attention to collective bargaining, working conditions, and direct organization of the workplace. The term business unionism is particularly useful because of its accuracy in describing the early unions' years of struggle to organize workers as well as to obtain contracts in the telephone industry in Canada.

Certainly, the early telecommunications unions may also be described as "permeable" because the first union to organize any workers in the industry was an international union, the International Brotherhood of Electrical Workers (IBEW), at BC Telephone. Jurisdiction for telephone workers was granted to the IBEW through its membership in the large Canadian central craft union the Trades and Labour Congress (TLC). But internal friction within the IBEW over the One Big Union Movement and the Winnipeg General Strike resulted in a weakening of the union at BC Telephone (Bernard, 1982: 65) that finally led to decertification. The only representation left to workers was a subservient company union, the BC Telephone Electrical Employees' Organization (EEO) (Bernard, 1982: 78). Having a company union during the Depression years helped BC Telephone because the EEO did not resist when the company laid off employees, reduced wages, and rolled back hours of work (Bernard, 1982: 81).

Although the IBEW's position among BC Telephone workers weakened after the Winnipeg General Strike, the union still continued to represent craft workers (installers) at AGT, MTS, NB Tel, MT&T, Island Tel, and Newfoundland Tel. The IBEW had obtained a charter in 1907 to organize the line workers at AGT. In 1917 the IBEW also organized the telephone craft workers and the operators at MTS (Manitoba Telephone System, 1990: 11). Unlike the locals in Alberta and British Columbia, the IBEW local at MTS was fairly militant, particularly in members' decision to join the other labourers in the Winnipeg General Strike. That militancy, however, was severely curbed during the 1920s and 1930s, as evidenced by a constitution that curtailed any union radicalism by requiring a 90 per cent strike vote and banned workers from taking part in any sympathetic strike action (Bernard, 1982: 98).

In 1907 Bell's aggressive management style was met with worker resistance when the Toronto telephone operators formed the Telephone Operators, Supervisors and Monitors Association and walked off the job (CEP, 1996: 5). By 1918 the International Brotherhood of Electrical Workers (IBEW) had signed up the association members. But Bell's vociferous anti-union attack on the IBEW resulted in its decertification three years later. The IBEW was subsequently replaced with a company union, the Canadian Telephone Employees Association (CTEA) (ibid., 5).[3] The impotent CTEA provided Bell Canada with a compliant work force, and its no-strike approach rewarded the company with eighty years of labour peace.

FORDIST LABOUR POLICY

By the mid-1940s most of the trade-union movement was incorporated into the Canadian Fordist "mode of regulation" (Jenson, 1989: 79). As part of Fordist regulation, the federal government passed the Industrial Disputes Investigation Act (IDIA) in 1948. Both the act and the IDIA board operated in a contradictory way by restricting and overseeing a union's right to strike. Furthermore, the act forced the union movement into business unionism by establishing only minimum standards for labour relations, forcing wage issues. Other labour policies included Order-in-Council Privy Council 1003, which established the right of private-sector employees to form and join unions; prohibited unfair labour practices; set up the machinery for defining bargaining-unit certification; mandated compulsory collective bargaining and conciliation; and established the right to strike once a collective agreement had expired (Panitch and Swartz, 1985: 13). In 1948, PC 1003 was superseded by the Industrial Relations Disputes Investigation Act, giving these rights a permanent legislative basis for private-sector workers such as those at BC Telephone and Bell Canada, who came under federal jurisdiction. The unions also won some security from the private sector through the Rand Formula, which authorized automatic dues check-off (Panitch and Swartz, 1985: 17).

For the most part, union organization of telephone workers occurred after these Fordist labour policies were introduced by the federal government. Once organization was completed, many of the unions representing telephone workers adopted "social unionist" practices.[4] Beginning in the late 1960s and continuing in the 1970s, 1980s, and 1990s, the social unionism practised by the Telecommunication Workers Union and the Communication Workers of Canada was far more aggressive and radical. For some of the unions, it meant a break from the international union and affiliation with Canadian national unions. Radicalism also grew as weak company unions were replaced with labour unions concerned with representing the telecommunications workers rather than company interests. Later in the 1980s, and 1990s political initiatives by the Telecommunication Workers Union and the Communication Workers of Canada included participation in telecommunications hearings at the CRTC and the initiation of extensive media campaigns to inform the public about the massive changes taking place in telecommunications.

The IDIA legislation was the catalyst that the EEO needed to transform itself (1943–44) from a company union to a legitimate union, the Federation of Telephone Workers Union (FTU). At the same time, the FTU obtained permission to affiliate with the central Canadian industrial union, the Canadian Labour Congress (CLC), rather than the central trades union, the Trades and Labour Congress (TLC). FTU's militancy began to blossom in the late 1950s when the new owners from the United States, the General Telephone and Electronics Corporation (GTE), drastically changed working conditions at BC Telephone.

Until the change in ownership, BC Telephone, like the other telephone companies in the post-war period, had undergone rapid capital expansion combined with technological upgrading. In order to finance its purchase, GTE cut BC Telephone's work force, introduced efficiency experts, intensified management's surveillance of workers, and increased the use of automation (Bernard, 1982: 137). By 1969 an explosive situation had developed, which culminated in the FTU's taking strike action over the issue of lay-offs and the increasing use of automation. The FTU's militancy increased as strikes and lockouts occurred throughout the 1970s over other issues such as management versus worker rights and technological change. The FTU's strike action ended fifty years of labour peace that had provided BC Telephone with enormous economic benefits. Throughout the 1970s the union's activism also grew as a result of the support and alliances it made with the militant British Columbia Federation of Labour and the CLC (Bernard, 1982: 131). By 1977 the FTU changed its name to Telecommunication Workers Union (TWU), not only to reflect the change in the industry from basic telephone service to telecommunications service but also to address the union's constitutional changes, which united the plant, traffic, and clerical workers. As a part of the Canadian national union movement of the 1970s the TWU revisited the nationalization issue, recommending that BC Telephone become a public institution. In the TWU's view, nationalization would prevent massive profits from going out of the country to the foreign-owned multinational.

In Saskatchewan in 1945, the social-democratic Co-operative Commonwealth Federation (CCF) was newly elected, and that year brought in the Saskatchewan Trade Union Act, helping to strengthen the position of unions in both the province and the whole country. The act helped to transform the Sask Tel's workers' association into an independent union, the United Telephone Workers of Canada (UTW) (CWC, 1984). Representing plant, traffic, and clerical workers, the UTW provides another good example of a permeable Fordist union. In 1950 the UTW merged with an international union, the Communication Workers of America (CWA). UTW's rationale for the merger was that the international union would provide better services and strengthen its bargaining position. None the less, it is important to point out that despite Sask Tel's public ownership, the corporation benefited, as did BC Telephone and AGT, from two compliant non-striking unions, the UTW and the CWA.

Canadian union nationalism grew in the 1960s and 1970s. The Canadian Communication Workers Council (CCWC), which had been established in 1967 within the CWA to run the affairs of the Canadian members, was keen to establish a Canadian national communications union. The council recommended that they break away from the international union, forming the Communication Workers of Canada (CWC) in 1972 (CEP, 1996; interview with Kinkaid, 1996). Under the direction of its new president, Fred Pomeroy, the CWC started out with 4,000 members, mostly from Sask Tel, a few workers from Northern Electric (now Nortel), and the plant workers at Toronto Telephone House. Within only a ten-year period the CWC grew from 4,000 to 20,000 members (CEP, 1996: 10, 16). The largest increase occurred in 1976, when the CWC was certified to represent the 12,000 maintenance, repair, installer, and line workers of Bell in Ontario and Quebec. What helped the CWC was a Canadian Labour Relations Board decision that forced Bell to allow union organizers to canvass on company property during non-working hours (ibid., 19–20). It was not until 1979 that the 7,400 mainly female operators and cafeteria workers switched to the CWC from the Communication Union of Canada (CUC), formerly the Traffic Employees Association (ibid., 20). Consistent with social unionism, CWC militancy grew as technicians and operators embarked on a series of strikes to achieve parity with western phone company employees and make reductions to the company's policy of mandatory overtime. Later, the CWC was also one of the first unions to obtain technological-change clauses in its collective agreements with Bell (Mosco and Zureik, 1987). But despite strong union representation at Bell, the CWC was unable to lure the clerical workers away from their company-sponsored association.

By 1975 a more militant IBEW at MT&T went on strike, for the first time in almost sixty years, over the issue of wage parity with other IBEW telephone workers in the west. The eight-week strike helped to sustain the union's new-found radicalism and to galvanize support for breaking from the international union. Shortly after the strike, an independent national union, the Atlantic Communication and Technical Workers Union (ACTWU), was established (interview with C. Simpson of ACTWU, 1996).

The shift in the Canadian telecommunications union movement from international to national union affiliation is indicative of the changes that were taking place throughout the 1970s across the Canadian labour movement. As Table 2.4 shows, from 1962 to 1992, Canadian union affiliation almost reversed. By 1992, 57 per cent of unions were represented by national unions compared to 29 per cent for the international unions. Moreover, the CEP emerged as the third largest national union, after the Canadian Union of Public Employees (CUPE) and the Canadian Auto Workers (CAW). CEP membership at 143,100 was greater than that of all the international unions except for the Food and Commercial Workers (CALURA, 1992: 36). Rather than build lasting alliances with the other telecommunications unions, the CEP has focused on increasing its union base through mergers with other non-telecommunication unions.

TELECOMMUNICATION UNION ALLIANCES

The failure to form a strong, lasting, cohesive alliance has been a major drawback for the telecommunications union movement in terms of presenting a united front to the telephone industry and the federal government. The first alliance was attempted in the mid-1940s, when the IBEW tried to persuade all Canadian telephone workers to affiliate with the TLC. But such an affiliation would have been problematic for the other unions because it would have excluded all their telephone operators and clerical workers. The FTU, by contrast, recognized that Canadian telephone workers would benefit from a national federation. The principles of the proposed federation provided the unions with lots of flexibility, including affiliation with either central trade union, the CCL or TLC; the retention of autonomy by member unions; and a commitment to standard agreements, covering working conditions, wages, and the promotion of good labour-management relations. Although the federation was thwarted by the IBEW and UTW, these principles provided the foundation for a subsequent telecommunications union federation. Recognizing the importance of co-ordinating the communication unions, the FTW and the CWC established their first formal alliance, the Canadian Federation of Communication Workers (CFCW), in 1973. The CFCW had a co-ordinating committee that was responsible for regulatory and legislative affairs. The CFCW also published a telecom-union newspaper and tried to organize all workers in the industry (interview

Table 2.4
Canadian affiliation with national and international unions, selected years

Year	1962 %	1972 %	1982 %	1992 %
International union affiliation	65	61	47	29
Nation union affiliation	22	26	36	57

Source: Adapted from Statistics Canada, *Corporations and Labour Unions Return Act* (CALURA) Cat. 71–202.

with J. Kinkaid of the CEP, 1996). Although the CFCW was created with the intent of merging the TWU and the CWC, thirteen years after it was formed the federation dissipated, as both unions chose to retain their autonomy. By the 1990s, however, all the telecommunications unions saw the need once again for an alliance that would share expertise and pool resources to combat the devastation to workers and telephone universality resulting from deregulation, liberalization, continentalization, and internationalization.

As this account of the Canadian telecommunications union movement reveals, for almost seventy years there was little effective resistance to the practices of the Canadian telephone industry. The various company unions sponsored by the telephone industry were extremely complacent about the exploitation of workers and did little more than legitimize the activities and practices of Bell and BC Telephone. The international unions, particularly the IBEW and the CWA, with their conservative approach did not consider strike action as a means of improving the working conditions or wages of telecommunications workers. Despite the introduction of somewhat favourable federal and provincial labour legislation, it would take another fifteen years before the FTW/TWU would develop into a militant labour movement. The FTW/TWU response to unreasonable demands and layoffs after the U.S. multinational GTE purchase of BC Telephone was a turning-point in terms of union radicalism. The second watershed occurred in the 1970s, with the rise of Canadian labour nationalism. The newly formed national CWC was both aggressive and militant in organizing the Bell Canada workers. Using strike action for the first time, the CWC dissented from and resisted the long period of Fordist telecommunications hegemonic consent.

FORDIST TELECOMMUNICATIONS POLICY

The economic crisis of the Depression necessitated a rethinking of the accountability of the federally regulated monopoly telecommunications providers. Beginning in the 1930s, the Canadian federal government began to take a more active role in formulating telecommunications policy. Most policy intervention was, however, the result of direct political intervention by the federal Cabinet through orders-in-council. By 1949 the Cabinet had put in place a statute to create a crown corporation, the Canadian Overseas Telecommunications Corporation (COTC) (renamed Teleglobe) (*Instant World*, 1971: 71). The COTC operated in Canada as an external telecommunications service for public overseas communications. This public-message telegraph service was a bold national endeavour that was made available by COTC and two foreign corporations, Western Union International and the Commercial Cable Company (a subsidiary of ITT World Communications).

In the early 1950s a mixed public/private microwave system was built to compete with land lines as an efficient transmission medium for data (record) and long-distance voice communications. As Babe notes, Bell Canada argued that only the Cabinet intervened, introducing broad policies to help develop a national microwave network and a TELEX system. Bell Canada argued that only the telephone industry should own the microwave system, excluding the telegraph companies. Cabinet maintained, however, "that no one person, or corporation, should receive monopoly to operate a microwave relay system ... [provided] the technical features of the applications submitted were satisfactory" (Babe, 1990: 129).

By 1958 the TCTC had a coast-to-coast system in place, while CNCP Communications concentrated their facilities east of Manitoba. Four years later the Cabinet stepped in once again, allowing CNCP to extend their system nation-wide. These Cabinet decisions helped to breathe new life into the old telecommunications supplier, CNCP, as well as to introduce competition into the telecommunications arena.

The Government Organization Act – 1969 created a new Ministry of Communication and the Department of Communication (Canada, S10 (1) (b)). According to the first minister of Communication, Eric Kierans, the prime motive of the Trudeau government in creating the department was to advance communications technologies by developing a satellite communications system. As a department then, DOC would promote economic and cultural benefits for regional development and provide better communications for the north (interview with E. Kierans, 30 Sept. 1996). Essentially, this federal department was created with the dual purpose of advancing capital accumulation for the resource sector in the northern frontier, for mining and gas exploration, and to promote and advance "the technological imperative" (DOC Annual Report 1977, 1978). Using this rationale, the federal state began to push the Canadian economy from its industrial/resource base into an information economy, where more goods and services would be produced by the growing information sector.

With a mandate that emphasized commercial, economic, and technical considerations, the Department of Communication (DOC) would "ensure the timely provision and optimum utilization of communications services and facilities in the national interest" (Canada, 1969, Appendix C, p. 2). DOC's functions included the formulation of broad national policies; regulating communications; planning and implementation of a national communications research progam; acting in an executive or advisory capacity to national and international telecommunications development progams; developing and recommending new regulation; and securing international arrangements for Canadian rights and interests in communications matters (ibid.).

Throughout the 1970s the new DOC embarked on a number of projects that required direct public financing for research and development at the DOC research institute, the Communication Research Centre (CRC). Engaging in applied research, the CRC and the DOC were both responsible for investments in an experimental satellite, Hermes, a direct satellite-to-home TV broadcasting system ($60 million) (DOC Annual Report 1976–77). Other research projects on coaxial cable benefited the newly emerging cable television industry, while research funding for fibre optics helped the telecommunications manufacturing industry. Public funds were also provided for research on a communications/telematic system that would be useful to the banking and finance industry (DOC Annual Report 1976–77). Other major investments included Telidon, the failed national videotex service ($40 million) (Mosco, 1982: 80; Wilson, 1988: 13–48). In a partnership progam with Bell Canada and CN Telecommunications, DOC funded the Northern Communication Assistance Progamme with federal public funds ($9 million) to bring long-distance service to the Northwest Territories. Together Bell and CN were to invest a similar amount in capital and operating funds for local exchange equipment and the operation of long distance.

The same year the Department of Communication was formed (1969), the Telesat Canada Act was passed so that a domestic communications satellite system could be built. Kierans states that DOC and the Telesat Act were part of the government decision to push for a satellite system despite the reluctance of TCTS members to invest in what they considered a premature satellite system (Kierans interview by National Museum of Science and Technology, 1990). The act allowed the government to create a private-public corporation, Telesat Canada. This hybrid company was established with tripartite ownership among the government, the telecommunications carriers, and the public (individual investors), each holding one-third of the shares. But decision-making, power and control actually rested with the federal government, which held 50 per cent of the voting shares, and the president of Telesat, appointed by the government, who had one voting share, ensuring that control of the corporation rested with the federal state (Commons, *Debates* , 1968–69, 7495; *Instant World* , 1971: 72–3). TCTS members and Island Telephone, *Québec-Téléphone* , Ontario Northland Communications, and CNCP Railway held the other 49 per cent of the voting shares. Through the launching of two satellites, Anik I and II, a national communication-service distribution system was created to compete with the existing land-based telecommunications system. But Telesat's dual obligation to provide Canadian content and operate on a commercial basis would later lead to major difficulties between the telecommunications companies and DOC. The tension in this dual mandate culminated in the privatization of Telesat almost twenty years later, with controlling ownership going to the TCTS consortium despite the enormous investment made by the public sector.

In the 1970s the Department of Communication introduced two telecommunications bills, C-43 and C-16, intended to establish a clear national telecommunications policy. The policy objectives stated that "efficient telecommunications systems are essential to sovereignty and integrity of Canada, and telecommunications services and productive resources should be developed and administered so as to safeguard, enrich and strengthen the cultural, political, social and economic fabric of Canada" (DOC Annual Report 1976–77: 1).

Although the proposed legislation focused on the development of broad national telecommunications policy, it also gave the federal government the right to direct broad telecommunications policies to the CRTC. At the same time, direct political interference in broadcasting policy, licensing, content programming, and quality standards, and restrictions on freedom of expression were off limits. After two successive attempts to introduce the legislation in Parliament, both bills died because they did not get the support they needed from the provincial governments. It took almost another twenty years before DOC was successful in advancing a revised telecommunications bill, which finally received royal assent in 1992, establishing the Telecommunication Act. Although some aspects of Bills C-16 and C-43 are reflected in the act, there are three important differences between the act and Bills C-16 and C-43: the erosion of national sovereignty, the entrenchment of economic rights over social and cultural rights, and the Canadian version of neo-regulation, forbearance, which is discussed later.

Of the numerous studies and conferences that DOC engaged in throughout the 1970s, the most significant was the Telecommission Studies, which were summarized in *Instant World* (1971). *Instant World* is an important document because of its declaration that the federal government has a responsibility to link telecommunications services with the public interest. The report explained that the state is responsible for the identification and protection of public interest in telecommunications by paying attention to the regulation of rates for telecom services (1971: viii). The report points out that, "where services are supplied in monopolistic conditions, the interests of the public must be protected. In Canada, the right to give this protection to users of telecommunications services is shared between the Governments of Canada and of the Provinces" (1971: 185). Concern was voiced over DOC's telecom policy objective that promoted new telecommunications technologies because of the danger that the nature and pace of development would be dictated by powerful private interests, such as the Telecom Canada members, and particularly the largest players, such as Bell and BC Telephone.

Instant World strongly recommended that the telecommunications sections of the Railway Act that dealt with just and reasonable rates (sections 320, 321) needed to be expanded to incorporate the principle of universality to public telephone service. But despite this concern for the public interest, the very structure and functions of DOC, with its emphasis on economic, commercial, and technological relations, contradicted the social importance of public interest. Even the CRTC, charged with the stewardship of service universality and affordability, continually interpreted these principles in a way that linked the public interest to economic objectives. From the late 1970s on, new telecommunications providers and large telecommunications users (primarily multinationals) organized into lobby organizations, mounting a powerful campaign that included lobbying the DOC, the CRTC, and other federal departments, to end the era of regulated telecommunications monopoly. According to the telecommunications users, their needs could only be served in a deregulated, privatized, liberalized environment. Consequently, domestic telecommunications policy would have to change so that these multinationals could participate in the expanding competitive global economy.

CANADIAN FORDIST
TELECOMMUNICATIONS REGULATION

Throughout the post–Second World War period, telecommunications was identified as increasingly important. This is why the federal state took a hyperactive role in promoting and investing in this sector. The objectives of the Second National Policy became the vehicle the federal government used to invest vast amounts of public funds in communications research and alternative telecom systems. In some cases public investment occurred in partnerships with the private telecommunications sector, contributing to the mixed public-private telecommunications system. But by the 1970s multinational domestic and foreign capital mounted a massive campaign to end the telecommunications monopolies.

From 1938 to 1967, under the mandate of the Department of Transportation, what little federal regulation there was of the telecommunications industry was conducted by the Board of Transport Commissioners (BTC). One important decision the BTC made was to adopt a new formula, "rate of return on rate base," which was, incidently, put forward by Bell. Under rate of return, a company is allowed to earn revenues to cover expenses plus a percentage return on investment (the base rate). Prior to this decision, rate regulation was based on a company's earnings per share. Rate-of-return regulation has always been considered controversial because it provided the telephone companies with built-in incentives for over-capitalization. In addition, there is always the possibility that rate of return can lead to predatory pricing in competitive markets. Forty years later, the telephone companies and the new telecommunications competitors would argue that the rate-of-return formula was outdated for a competitive environment and should be replaced with "cost-base pricing."

In 1967 telecommunications regulation was transferred to a new regulatory body, the Canadian Transport Commission (CTC). The CTC, however, with its formal panel of judges, was not structured to be able to deal with the pressure from new technologies such as microwave, satellite, or computer-telephony integration (Coulter 1992: 173). Under a judiciary system the CTC could do little to address the mounting complaints and growing conflicts between the established telecom monopoly service providers and the new telecommunications competitors. No doubt the CTC's outright hostility towards the new telecommunications providers, and growing complaints from public-interest groups, helped, along with recommendations from DOC and *Instant World*, to transfer telecommunications regulation to the communication ministry. Regulation "in the public interest" took place through the more policy-oriented regulatory body, the Canadian Radio-television and Telecommunications Commission (CRTC).

As a semi-autonomous regulatory agency the CRTC ostensibly operates as a pluralist institution that, in conjunction with Parliament, the cabinet, the federal and Supreme courts, and the Departments of Industry and Heritage, provides the regulatory rules and procedures for the federally regulated telecommunications common carriers. The Supreme Court, the Federal Court of Canada, and the CRTC are then responsible for telecommunications process decisions. Under the CRTC, telecommunications regulation was finally opened to the public, accompanied by notices of upcoming hearings that anyone can attend (CRTC Annual

Report 1979–80: 2, 3). Nevertheless, despite the establishment of public hearings that broadened participation to include the telecom suppliers, new competitors, business users, telecommunications unions, various public-interest groups, and other consumers, it is important to keep in mind that the regulator (CRTC) is part of the state apparatus and an "unequal structure of representation" (Mahon, 1980: 157). Although the CRTC allows for public participation in the telecommunications decision-making process, what it really does is to redefine the participants as individual subjects. This has two effects: first, telecommunications decisions tend to promote consumerism and individual telecom consumption. Second, inequality is reinforced, because the regulatory process restores the position of the established and new telecommunications service providers, while the rest of the participants, the unions, public-interest groups, and the communications consumer groups are left fragmented and often divided on policy issues.

3 Telecommunications Liberalization: Phase One

Hegemony is also embodied in a range of substantive ideas such as the widespread acceptance of the marketplace as the cornerstone of a productive economy, of voting as a primary means of carrying out democracy, and of journalistic objectivity as the product of two views on an issue of the day. These and other hegemonic views (free markets, free elections, a free press, the free flow of information, etc.) are neither politically neutral values, nor ideological instruments of control imposed from above.

Vincent Mosco,
The Political Economy of Communication

After the economic crisis of the 1970s, transnational corporations operating in Canada intensified their pressure on the state to reorganize macro and micro policies. Changing Canadian policies to implement a neo-liberal model would require advancing market interests to the policy forefront. However, support for such policies would require the production of hegemonic consent among the various economic sectors. Key to producing such consent for this policy shift were actors from within the federal government and Canadian and U.S. transnational corporations and their lobby organizations, as well as experts and academics funded by both the private sector and the state. A reaction against hegemonic consent was met with strong opposition and resistance, which will be discussed in greater detail later. The manufacture of consent and resistance to the neo-liberal reorganization of telecommunications policy required a dynamic interplay between the federal government and business, as well as opposition from labour and other social groups. It is in this way that hegemony is not simply imposed by class or state power but is organically constituted and embedded in the social and political relations involved in the policy process. Normally hegemony involves compromises and trade-offs, usually among businesses, the state, workers, and other social groups. We need to keep in mind that at one level businesses do compete with each other, but at another they tend to co-operate to advance their common capital interests. Moreover, the implementation of a neo-liberal policy approach also tends to intensify lobbying by many corporations and their various associations and coalitions; essentially, this lobbying intensifies capital political advocacy.[1]

The neo-liberal project included reducing the role of the state in the economy through policy changes that included liberalization, neo-regulation, privatization, continentalization, and internationalization. Key sectors of the economy, such as banking, transportation, and telecommunications, were targeted as extremely important for the multinational capital-accumulation project and in need of policy reorganization. The neo-liberal project also required the reorganization of the Canadian telecommunications sector so that state power could be centralized, bringing all telecommunications under federal jurisdiction and regulation.

Neo-liberal philosophy helped to reform the policy agenda so that individual rights were favoured over collective rights, and in the place of state regulation, the free market was restored, as was market regulation. For Canada this policy shift also included replacing the Second National Policy with an overarching continental-trade agreement/policy. The new agenda would combine a free-market approach with free trade between Canada and the United States, which was extended to include Mexico.

Changing the telecommunications environment from one of regulated monopoly to an ostensibly open competitive market helped to integrate the country into the continental telecom regime. Reorganizing business and creating public consent for a neo-liberal policy approach was not so simple. The telephone companies and the majority of telephone subscribers were not as easily convinced that this was the direction that Canadian telecommunications policy should go.

MULTINATIONAL LOBBYISTS

According to William Stanbury's *Business Government Relations in Canada*, the capitalist class devised a strategy that would apply direct and indirect pressure on the Canadian state to change public policy (1993: 240). Direct techniques included forming alliances and coalitions; participating in formal and informal meetings with policy-makers (bureaucrats and politicians) and their advisers; submitting briefs to departments; and providing testimony before parliamentary committees to advise formally on policy and legislative issues. Indirect techniques included the funding of think-tanks (research institutes) to try to change the composition of the set of ideas around public policy; the generation of new information and analysis between public and private policy actors; and advertising to sell business ideas to the public and policy-makers. These broader business alliances helped to create an integrated web of hegemonic consent[3] particularly among both domestic and foreign multinational corporations in their attempt to influence the federal government and to convince the Canadian public to accept the neo-liberal philosophy.

Big and small continental business coalitions were formed to promote the neo-liberal project. Reorganizing hegemonic consent for the project occurred as organizations such as the Business Council on National Issues (BCNI) and the Canadian Federation of Independent Business (CFIB), among others, promoted a continental business agenda. At the same time the BCNI and CFIB promoted ideas to rein in a federal government that had grown too intrusive into the Canadian economy. Formed in 1976 as the umbrella organization for big business, the BCNI brought multinational and conglomerate capital together on common issues. Organized as a political-policy actor, the BCNI lobbied the federal government for neo-liberal policy changes. One hundred and fifty chief executive officers from the largest corporations operating in Canada make up BCNI's exclusive membership (see Appendix 1, Policy Committee). Modelled on the American Business Roundtable, the BCNI was designed to develop a common front for multinational capital by advancing its common interests. These interests include the free flow of capital, goods, and services, and a liberalized trade and investment environment so that its business members could compete in an integrated and globalized marketplace (BCNI, 1993: 20). As a new business coalition, BCNI's approach was broader and went beyond the sectoral interest of other associations such as the Canadian Bankers Association, the Canadian Manufacturers Association, the Canadian Chamber of Commerce, and the Canadian Petroleum Association.

BCNI membership integrates all the key sectors of the Canadian economy: resources, manufacturing, finance, transportation, utilities, services, and the communications sector (see Table 3.1). The council is truly continentalist because it integrates Canadian and foreign (mostly United States) multinationals. Throughout the 1980s the primary concern for BCNI members was gaining unhampered access to the United States and Canadian markets, as well as the international market. Main interests included increasing competitiveness in the national economy, improvements in international economy and trade, and institutional government reform. These interests are best summed up in BCNI's statement that its purpose was to "strengthen the voice of business on issues of national importance and put forward constructive courses of action for the country" (cited in Langille, 1987: 48).

While there are many competing capital interests among BCNI members, its broad objectives were, and continue to be, defined by the BCNI executive and policy committee, where internal decisions are arrived at through compromise and consensus (A. Sinclair, BCNI, February 1996).

Table 3.1
BCNI Sectoral Membership

Sector	Members
Business lobby associations	4
Communications	12
Finance	32
Manufacturing	65
Resources	9
Services	11
Transportation and utilities	10
Wholesale and retail	7
Total	150

Source: Adapted from Business Council on
National Issues, 1994.

Cross-integration among BCNI members is perpetuated through sectoral membership in associations such as the Canadian Bankers Association, the Canadian Manufacturers Association, the Canadian Chamber of Commerce, and the Information Technology Association of Canada (see Table 3.2). Cross-continental integration is also achieved through membership in the Canadian American Committee and the Trilateral Commission. This integrative aspect continues through members who are also board members of private research organizations such as the international Conference Board of Canada[2] and the C.D. Howe Institute. A number of members also sit on important federal government committees such as the Board of Trade, the International Trade Advisory Committee (ITAC) and the Sectoral Advisory Groups on International Trade (SAGIT). Many of BCNI's CEOs also sit on the boards of directors of other members' companies and hold significant minority shares in each other's companies, and/or jointly hold significant shareholdings in other dominate multinationals (Richardson, 1992; Coulter, 1992). In addition, the finance-capital sector (see Appendix 3) provides an important "intermediation" function across all the other sectors, for both domestic and international capital (Reddick, 1993: 51). Such a dense network strengthens the position of multinational capital, creating an integrative hegemonic web of consent that was bent on neo-liberal policy reform.

As a distinct sector, communications' importance is evidenced by the BCNI members involved in the communications sector (see Table 3.3). Over the last few years the communications sector has gained importance for Canadian capital and the state because of its strategic importance to all the other sectors, particularly multinational capital. Because multinational corporations rely on new technological developments such as computers, telecommunications, and satellites for

Table 3.2
BCNI Executive and Policy Committee cross-affiliations

Association	Number of BCNI members
CANADIAN ASSOCIATIONS	
Calgary Petroleum Club	1
Canada West Foundation	2
Canadian Bankers Association	3
Canadian Manufacturers Association	2
Canadian Mining Association	1
Canadian Chamber of Commerce	4
Information Technology Association of Canada	1
CONTINENTAL AND INTERNATIONAL ASSOCIATIONS	
Asia Pacific Foundation of Canada	1
British North American Committee	1
Canadian American Committee	1
International Business Council of Canada	1
Trilateral Commission	3
World Economic Forum	1
PRIVATE RESEARCH ORGANIZATIONS	
Conference Board of Canada (U.S. subsidiary)	9
C.D. Howe Institute	5
Van Horne Institute	1
CANADIAN FEDERAL GOVERNMENT COMMITTEES	
Board of Trade	2
International Trade Advisory Committee (ITAC)	3
Sectoral Advisory Groups on International Trade (SAGIT)	2

Source: Adapted from Who's Who in Canadian Business, 1996.

production, distribution, and other exchange- and trade-related activities, multinationals, especially those in the communications sector, have brought these technologies together to form an integrated "global grid" of communications networks (Schiller, 1985: 105).

THE INFLUENCE OF RESEARCH INSTITUTES

Neo-liberal ideas were also advanced through business funding of private research organizations and state funding of public research institutes. Sympathetic organic intellectuals,[3] by and large the neo-classical economists and lawyers, among others, all helped to contribute to the diffusion of these ideas. Often these organic intellectuals worked for both the private sector and the federal government and provided

Table 3.3
BCNI communications membership

Businesses	1986	1994–95
	Member	Member
TELECOMMUNICATIONS		
Bell Canada Enterprises Inc.	x	x
British Columbia Telephone Co.	x	x
Maritime Telegraph and Telephone		x
New Brunswick Telephone Co.		x
Northern Telecom Ltd.	x	x
Teleglobe Canada		x
Unitel Communications		x
MEDIA COMMUNICATIONS		
Reader's Digest Canada		x
Rogers Communications		x
Southam Inc.	x	x
CTV Television Network		x
Maclean Hunter (Rogers Communications)		x

Source: Adapted from Business Council on National Issues, 1986, 1994–95.

evidence at regulatory hearings, as well as advice and testimony before royal commissions or parliamentary committees.

Privately funded research institutes have been instrumental in promoting the neo-liberal agenda. It is not surprising that neo-liberal ideas have diffused to privately-funded research institutes such as the C.D. Howe or Fraser institutes, because their membership and funding come primarily from big business (see Appendices 4 and 5). As early as the mid-1970s the C.D. Howe Institute was producing research that endorsed the reduction of government intervention in private economic activity. In the institute's view, government regulation and regulatory bodies had been necessary in the past to prevent private market failure, or the adverse effects of natural monopolies. But they were no longer necessary in the changing environment of free markets and free international trade. As a result, the institute produced research that endorsed deregulation as a way to address the state's overinvolvement in the Canadian economy (C.D. Howe, 1976: 18).

The institute also saw communications emerging as a strategic issue for Canadian public policy. Recommendations for competition in the communications industry were to be combined with policy liberalization, particularly for the telephone carriers (C.D. Howe, 1980: 69–73). The centralization of telecommunications jurisdiction and regulation provided a way of addressing the split levels of regulation among federal, provincial, and municipal governments. According to the C.D. Howe Institute, the growing importance of telecommunications policy required a comprehensive and integrated treatment, where by the existing regulatory patchwork would be centralized to bring telecommunications under federal jurisdiction and regulation (C.D. Howe, 1980: 7, 34).

A number of BCNI members are also on the board of another influential private research organization, the Fraser Institute. Advocating neo-liberal policies, the Fraser Institute has been vociferous in its attacks on public ownership, producing research to support privatization. Often the institute's reports offer abstract neo-classical ideal economic formulas and a pro-market dogma as policy recommendations. This research takes on a fundamentalist, religious-like fervour, particularly in its reliance on Nobel laureate Milton Friedman's recommendations on how to privatize state holdings. Privatization, in the Fraser Institute's view, is the policy alternative to the many forms of government intervention in the marketplace. Private entrepreneurs are preferable to public ownership because of capitalism's property rights: the right of ownership; the right to control; and the right to receive profits or bear losses (Fraser Institute, 1980: 158). This tautological reasoning supports privatization because private business, driven by self-interest, will always seek what is best for business, which in turn leads to what is best for society (Fraser Institute, 1980: 158–60). Consequently, it is not surprising that the Fraser Institute recommended that Teleglobe and Telesat be privatized so that consumers could benefit from access to more services and greater choices in consumer goods at lower prices (1980: 31, 153–4).

POLICY ACTORS WITHIN THE FEDERAL GOVERNMENT

Two publicly funded research organizations, the Economic Council of Canada and the Institute for Research on Public Policy, both endorsed the neo-liberal agenda. In its annual review, *Lean Times*, the Economic Council stated that the federal government needed to reduce its regulation of the economy because it impeded the free operation of the market (1982: 107). The council recommended that a number of federal regulatory bodies, such as the agricultural marketing boards and those dealing with the airlines, the environment, and telecommunications, should reorganize. Specific recommendations directed at reforming telecom regulation included opening up entry to other telecommunications service providers who specialized in private-line and data-communication services; allowing the interconnection of CNCP to the incumbent telecommunications companies; and permitting competition in all forms of telecommunications equipment. Other recommendations called for lifting all restrictions on leased lines, as well as the resale of . long-based transmission facilities (Economic Council, 1981: 88).

The Institute for Research for Public Policy (IRPP, Appendix 8), in addition to endorsing regulatory reform through deregulation, supported the privatization of as many of Canada's Crown corporations as possible. In a jointly sponsored conference the IRPP, the Bank of Nova Scotia, and McLeod, Young and Weir provided their strategy and recommendations in *Papers on Privatization*. Sympathetic intellectuals such as Thomas Courchene (1985: 2) and Richard Lipsey (1985: 37) both claimed that private corporations are far more efficient than Crown corporations. Research produced for the IRPP came to a common agreement that public enterprises and government operations needed to be opened up to market forces. Privatization policies, then, would be one way, among others, to strengthen existing market forces and open up Canada's mixed economy.

In seeking a solution to the economic crises of the 1970s, the Canadian state established a royal commission to inquire into the country's perceived economic and political problems. The Royal Commission on the Economic Union and Development Prospects for Canada, or the Macdonald Commission, as it is commonly known, provided the federal impetus to reform Canada's economic and social policies. It was thought that the Macdonald Commission would contribute to shaping public debate and creating consensus on divisive issues that would result as the state moved Canadian society to an "open-market economy." According to Drache and Cameron, royal commissions are often used in Canada to defuse explosive issues, and the Macdonald Commission was no exception (1985: xi). As part of the hegemonic process to reorganize consent for continentalization and internationalization, the commission claimed that its investigation would be impartial. This would lead one to believe that the MacDonald Commission's public-forum process was specifically used to lay the groundwork for building social consent for neo-liberal economic policies. To make sure that the neo-liberal view dominated, only the perspective of neo-classical economists shaped this commission (McFarlane, 1992: 293). McFarlane notes that by the mid-1980s, royal commissions were out of favour with the federal government, which had come to prefer internally handled task forces because their findings were not public, as were those of the royal commissions (293). The choice to hold a public forum rather than a task force appears to have been a deliberate one in that the Macdonald Commission also helped to garner hegemonic consent for neo-liberal policy continentalization.

The commission's interim report, *Challenges and Choices: A Commission on Canada's Future* (1984), provides insight into the reasons the commission was set up. Concerns had been raised that trade issues such as heightened competition in domestic and world markets, the growing internationalization of business, the effects of technology on trade, and the growing protectionism of Canada's largest trading partner, the United States, would leave the Canadian economy trailing in the wake of these catastrophic pressures (Macdonald Commission, 1984: 11). The report stated that Canada needed to re-examine its policies to ensure that the country was aware of the implications of these changes and others, such as the emergence of trading blocks (the European Common Market), the movement of capital between countries, the changing role of resources (with particular implications for Canada), and the availability of large labour pools in developing countries. It was further noted that the Canadian state should respond with appropriate policy changes. According to Simeon, "the choice ultimately was the market model or nothing" (1987: 177) because "there was no credible alternative before the Commissioners" (173). A closer examination of the commission's findings and recommendations, however, reveals that three different policy visions were put forward.

The first view came from the business sector, including the BCNI, the Canadian Manufacturers' Association, the Canadian Bankers Association, the C.D. Howe and the Fraser institutes, neo-classical economists, and many of the provinces, including those in the west, Ontario, and Quebec. Simply stated, this view held that the scope of the Canadian government should be reduced, particularly government intervention into the economy. There should also be a major reduction in state public ownership. Government policy instruments would have to be more market-driven to facilitate the necessary adjustments to expand international business. This view also held that Canada should establish a long-term trade strategy and complementary policies with its major trading partner, the United States.

The second view, from government agencies such as the Science Council of Canada, envisaged a new mandate for the state and an important role for Canada's high-tech industries as a partial measure to move the country towards an information economy (Science Council of Canada, 1982). This view called for the creation of a new social accord between workers and their employers, similar to the European corporatist model, a proposal that was, however, rejected by both business and labour.

A third view, for sustainable economic development that would ensure social as well as economic security for all citizens, was put forward by ordinary Canadians from what has been referred to as the "popular sector" or the popular forces (Drache and Cameron, 1985: ix).

Adopting an elite model of inquiry, the commission conducted numerous public hearings and meetings. National debates, expert opinions, discussion papers, and individual testimony were presented at hearings held in twenty-seven communities across the country. Presentations were made by 1,827 interveners, and 1,100 written submissions came from individuals, organizations, and governments (Macdonald Commission, 1985, Appendix 6). At first glance it would appear that the commission went to great lengths to consult not only business interests but academics, labour, citizen groups, and the public at large. But despite conflicting views, the commission endorsed and adopted the pro-market view and a neo-liberal policy model. The function of the commission then was to garner hegemonic consensus to help to legitimize the state shift to a neo-liberal policy environment. As the commission stated in volume 1, "this is a time for reassessment, for the search for new directions, for the application of analysis unencumbered by an automatic acceptance of yesterday's assumptions. Such an assessment is not easy ... In general, we seek, in the interest of efficiency, to modify the network of policy links connecting the state and the economy and society of Canada. This requires governments to reduce the disincentives which their own policies have placed in the way of behaviour that fosters an adaptive economy and society."

ALTERNATIVE ACTION

Strong resistance to the continental regime developed none the less. Opposition or alternative hegemony developed over a number of neo-liberal policy initiatives and to both free-trade agreements. Political alliances formed between class and progressive social movements. Alliances developed within the working class, including among both the private- and public-sector trade unions, and between the unions and the other social groups and organizations. The first broad alliance was called the Pro-Canada Network. This alliance changed its name to Action Canada and widened its mandate from national issues to continental and international concerns. In addition, the alliance built continental and international links with other progressive social groups and trade unions in Mexico and the United States. The membership of Action Canada included cross-denominational church organizations (GAT-FLY, renamed the Ecumenical Coalition for Economic Justice), social agents, native peoples' organizations, women's organizations, OXFAM Canada, the Canadian Federation of Students, and InterPares, among others (Bleyer, 1992: 104–14). Individual organizations as well as the alliance presented counter-briefs and lobbied against the Macdonald Commission's recommendations. And, through alternative research organizations such as the Canadian Centre for Policy Alternatives, other intellectuals provided research and a number of reports on the negative impact that a shift to a neo-liberal policy model and continental free trade would have on Canadian society.

The extensive fanfare of the Macdonald Commission did not, however, stop progressive organizations from presenting their alternative views. The public sector, with representatives from labour, community groups, social agencies, churches, farmers, volunteer organizations, women's and youth organizations, the unemployed, native groups, and the regionally disadvantaged, presented the commission with their vision for Canada (see Appendix 7).

Concerned about the relationship between social and economic policy, these representatives of the public sector reiterated their views in *The Other Macdonald Report*. In numerous submissions at the first round of public hearings, common themes emerged concerning social and economic security for all Canadians. These themes included sustainable economic development based on strategic planning and constructive internationalism (1985: xxx–xxxix). According to the public sector, Canada did not have a productivity problem or an unfavourable investment climate. Rather, the Canadian government had an incoherent industrial strategy and an employment crisis (Social Planning Council of Metropolitan Toronto, 1985: 21–2; United [now Canadian] Auto Workers, 1985: 23–4; Ontario Public Service Employees, 1985: 136–45). The cornerstone for any industrial strategy, according to the public sector, is the expansion of public services (such as education, day care, medical services) to create employment and improve the lives of all members of Canadian society. Other common themes included expanding the role of government in the economy and new forms of government intervention based on grass-roots involvement (democratic participation) in government planning (United Church of Canada, 1985: 183). It was also emphasized that social justice in Canadian society requires fairness and equitable treatment of the poor and the marginalized (United Church, 1985: 183; Canadian Conference of Catholic bishops, 1985: 213). The Catholic Bishops also held the view that the Canadian government lacked the political will and moral vision to develop an alternative to the market model that was being advanced (1985: 213).

Other views were put forward regarding the importance of maintaining public ownership in certain areas of the economy. It was thought that public ownership would break the stranglehold of the multinationals and ensure public accountability and some public control over economic policy (CUPE, 1985: 196). The alternative research focused on the problem of placing too much reliance on new technologies, as well as the so-called information economy. As the UAW explained, high technology, embedded with computer chips, had already been integrated into many sectors such as manufacturing, services, and resources. Furthermore, increasing evidence showed that the so-called technological revolution, as it extended to employees in the workplace, not only enhanced some work and created some jobs, but disproportionately deskilled work and displaced jobs (OPSEU, 1985: 144).

These alternative views made two important contributions. First, they identified and criticized the short-term interests and demands of business. They charged that business interests and the market are embedded with a particular bias that is inappropriate for establishing public policies because they exclude other important social and cultural considerations. Second, the common ground reached by these alternative political actors helped to serve as the seed-bed of the more developed coalitions that formed in the 1980s and 1990s. Not only did these dissenting voices include subordinate class actors; they also cut across class boundaries, creating what Mosco calls an "alternative hegemony" (1996: 243) that continued to oppose the multinational capital agenda and the neo-liberal policy shift. Resistance continued to build in sectors such as telecommunications as the federal government began to implement the neo-liberal model. As the following chapters reveal, these telecommunications policy changes were met with continuous opposition.

4 External and Internal Pressures on Canadian Telecom Policy Reform

Canadian telecommunication policy ... reflects the demands arising from new realities, including the globalization of markets, in particular the integration of the North American market, and the extremely rapid development of the telecommunication Sector.

Communications Canada, 1992

EXTERNAL PRESSURES FROM THE UNITED STATES

The analysis of Fordism as developed by Gramsci (1971) and further elaborated by the French regulation school explains how a Fordist regime of accumulation and regulation works (Aglietta, 1979; Lipietz, 1987). The Fordist regime that was put in place by the American state to counter the economic crisis of 1929 included production processes, welfare-state macroeconomic policies, and compromises with labour and American citizens. Relying in part on technological innovation, the U.S. Fordist regime benefited from production improvements and labour's acceptance of technological change. Social compromises made by business and the state with labour included labour legislation, collective bargaining, and wage increases for unionized workers. Economic growth in the post–Second World War period provided increases in income not only for unionized workers but for the rest of the labour force. Improvements in wages led to more disposable income and increases in the consumption patterns of American-produced goods and services.

This Fordist regime also had an impact on U.S. telecommunications service production, employment, and policy. The Ozark Plan was one part of American telecommunications Fordism (Bolton, 1993: 123; Horwitz, 1989: 235). Commencing in the early 1950s and continuing until 1980, when the Federal Communication Commssion (FCC) allowed competition in long-distance telephone service (message toll service – MTS), the plan promoted the expansion of local telephone service with the aim of "putting a telephone in every home" (cited in Bolton, 1993: 124). Successful because of its labour inclusiveness, the Ozark compromise was based on a loose consensus among AT&T and its twenty-two regional Bell companies, the involvement of major unions such as the IBEW and the Communication Workers of America (CWA), and the consumers (telephone subscribers). The telecommunications service sector also relied on regulation that was divided between the FCC, for intrastate activity, and the Public Utility Commissions (PUCS), for interstate activity. This policy environment not only helped to increase the telecommunications work force but also provided relative labour peace for AT&T, in much the same way that the Fordist telecom regime did in Canada.

The post-war capital-expansion progam helped further to develop and regenerate the U.S. telephone network as well as the entire telecommunications industry. American telecom Fordism was aided by the Ozark system, which linked AT&T's subsidiary manufacturing operations, Western Electric, with its Bell Laboratory research operations, to its local and toll telephone services. AT&T benefited through its vertically integrated operations because it was the sole supplier of telecommunications equipment to its long-distance operations and the regional Bell companies. Through FCC and PUC regulation the capital-expansion project was tied to company rate increases and to specific standards regarding quality of service. The state oversaw a system in which AT&T's position was defined and regulated as a natural telecommunications monopoly. AT&T also benefited from increases in telephone penetration levels, as Table 4.1 indicates, because state regulators linked penetration to universal access to telephone services. As the standard of living went up from the 1950s to the 1980s, telephone customers responded by putting more phones in their residences, increasing the number of local and long-distance calls they made, which resulted in a penetration rate of 62 per cent in 1950, and increased to 96 per cent by 1980. As noted by Horwitz (1989: 126), the FCC and the PUCS provided a long period of regulatory normalcy and consistent basic regulatory procedures, formula, and policies. AT&T was permitted to exercise monopoly control over long-distance and local phone service. Rates were approved when requested through fair rate of return. This helped AT&T to stabilize the cost of doing business and kept basic local phone rates low.

Telephone services expanded in this period as mandated by Section 1 of the Communications Act. The act, FCC policies, and PUC regulation helped telecom capital expansion while subjecting the industry to some public controls that in theory were tied to the social interest, in the

Table 4.1
u.s. telephone statistics 1950–80

	1950	1980	Increase (%)
Residential and business phones	39 million	157 million	402
Number of local calls	65 billion	279 billion	429
Number of long distance calls	2 billion	32 billion	1600
Penetration rates	62%	93%	150

Source: FCC, adapted from Statistics of Common Carriers, 1960, 1980, pp. 21, 18; FCC,
Annual Report, 1960, 1980.

form of fair rates. Customers were pleased with the plan and tended
to believe in and support the idea of monopoly telephone service.
Ostensibly, monopoly regulation permitted an extensive system of
subsidization where local rates were kept low for subscribers through
long-distance revenues. Local residential rates were also subsidized by
business subscribers and the cost of rural services was kept low through
a subsidy from the densely populated urban serviced areas.

Through legislation and regulation both federal and state levels of
government played an important role in securing labour in the Fordist
policy compromise. As an example, Congress amended Section 222(f)
of the Communications Act in 1943 to require the FCC to evaluate the
effect of mergers and acquisitions on employment. Moreover, as early
as 1910, the Department of Labour took an interest in the conditions
of telephone workers and their jobs (Winseck, 1993: 123). And until
divesture, as Winseck (123) explains, a historical precedent had been
established by the legislators and regulations always to consider policy
effects and decisions on telecommunications labour.

The telecommunications trade-union movement was also an important part of this Fordist compromise. With considerable growth in
telecommunications employment, union membership also increased.
Along with the Bell companies' no-lay-off policy, that growth gave the
telecom unions a considerable stake in the industry. Table 4.2 shows
the composition of telephone industry employment from 1950 to
1980. Through this period total employment increased from 560,700
in 1950 to 903,700 by 1980. By 1980, 80 per cent of the major
telephone-company employees were union members (ILO, Bolton,
1993: 132). Additionally, the trade unions accepted, and were even
advocates of technological change. CWA union contracts were some of
the first in North America to include technological-change clauses.
Nonetheless, this devotion to telecommunications technology backfired
after the unions accepted regulatory reform and competition in customers' telephone equipment and public long-distance service.

Table 4.2
Composition of u.s. telephone industry employment 1950–80

	1950	1960	1970	1980
Managerial & professional	57,600	92,900	151,700	211,300
Clerical & operators	345,600	325,500	394,000	332,200
Engineers & technicians	133,200	175,000	258,000	329,700
Other	24,300	26,200	27,400	30,500
Total	560,700	619,500	831,600	903,700

Source: ILO (1993), adapted from *Telecommunications Services*, 130.

The u.s. Fordist regime came under attack from a number of forces, including business, business lobby organizations, and the federal government. As support shifted in favour of a neo-Fordist regime, a reorganization of the previous hegemonic project was required. Multinational corporations wanted a more flexible regime that would be conducive to further domestic and international capital-accumulation expansion. Consequently, u.s. businesses responded to the capital crisis of the 1970s with a hegemonic project based on neo-liberal policy reform. To implement this agenda, individual corporations made new alliances and created new structures to co-ordinate with one another and to influence the federal government. The most significant organization, the Business Roundtable, formed in 1972 as a result of a merger of three ad hoc business committees, exerted political pressure and lobbied Congress for regulatory reform in sectors such as trucking, air transport, banking, and telecommunications (Horwitz, 1989: 206). Exclusive membership in the Roundtable consisted of the chief executive officers of major corporations in industry, resources, finance, commerce, and retail, among others. The 1979 Arthur Anderson & Co. report, the *Cost of Government Regulation Study*, critized all forms of government intervention as well as congressional liberals, union presidents, and former regulators.

Another new organization, the National Federation of Independent Business (NFIB), with 596,000 members, was formed with the sole purpose of exercising political influence at the federal level to reform regulation (Horwitz, 1989: 231). An older established trade organization, the u.s. Chamber of Commerce, under new leadership from Proctor and Gamble, u.s. Steel, and General Motors, was revamped into an aggressive political lobby organization. From 1974 to 1980 the Chamber's membership doubled, while its annual budget was substantially increased to $68,000,000 (Pertschuk, 1982: 57). Both the Chamber of Commerce and the NFIB were very critical of the existing regulatory environment, focusing on the inflexible aspects of regulation in trucking, the airline industry, banking, and telecommunications.

Alliances were also made between corporations and academic neo-classical economists. In the 1960s some of these economists conducted empirical studies on the regulation of the previously mentioned economic sectors. Major among these studies was an assessment of regulation and price, entry, and exit regulation. Findings published in influential economic journals such as the *Journal of Law and Economics* and the Bell *Journal of Economics and Management Science* (now the *Rand Journal of Economics and Management Science*) concluded that government regulation needed to be reduced and that the economy should rely instead on the forces of market competition (Horwitz, 1989: 206).

Neo-classical free-market economists took a number of key positions in the Ford, Carter, and Reagan administrations in the Department of Transport, the Council of Economic Advisors, and the Justice Department. Once inside the government, these economists promoted pro-competitive deregulation as a policy goal. By the late 1960s the anti-trust division of the Justice Department had begun to develop pro-competition regulatory reform within the government as a way of fulfilling its mission. By the early 1970s reform political advocates inside the executive, particularly in the Department of Justice and the Office of Management and Budget, dominated setting government information policies. Reformers in the State Department (responsible for international matters), the National Telecommunications and Information Administration (NTIA) (a Commerce Department agency created in 1978 and responsible for policy research and co-ordination), as well as the Federal Communications Commission were all committed to the goal of neo-liberalism (Derthick and Quirk, 1985: 37; Mosco, 1989: 184).

Another business strategy employed to influence the political agenda was the corporate funding of conservative foundations such as the Olin and Ford Foundations, and policy-research organizations such as the American Enterprise Institute for Public Policy Research, the Institute for Contemporary Studies, the International Institute for Economic Research, the Center for Study of American Business, the Hoover Institution, and the Brookings Institution (Derthick and Quirk, 1985: 36; Horwitz, 1989: 207). Research conducted by these institutes helped to foster opinion that the U.S. economy was overregulated. Regulation, according to these research organizations, was responsible for the dismal performance of the U.S. economy in the 1970s, its high inflation, low productivity, and the country's slow economic growth. Studies from these research organizations helped to bring about changes in the political discourse, which opened the way for regulatory reform. These same lobby organizations, economists, and research institutes found fault with the whole regulatory process. Increases in the number of claims and the number of participants at regulatory hearings in the 1960s and 1970s led these critics to latch on to the idea that regulation contributed to an "overloaded government" (Easton, 1965). Direct criticism of the existing telecommunications sector came from policy-planning groups with strong links to big business. The Conference Board, the Committee for Economic Development, and the Aspen Institute Program on Communications and Society, along with government officials within the u.s. executive departments, all supported the neo-liberal agenda and the move to reform telecommunications policy (Haight and Weinstein, 1981: 139).

U.S. TELECOM LIBERALIZATION AND NEO-REGULATION

New telecommunications needs in the United States arose within the larger context of multinational expansion in the post-war period. Large telecom users such as multinationals and domestic conglomerates saw the benefit in new technologies and new delivery systems such as microwave and satellite networks, computer technology and digitalization, and fibre-optics cable over the rigid coaxial-cable land-based system offered by AT&T. According to the large users, the new telecom-network providers, and the new communications-equipment manufacturers, these alternative systems were more cost-effective, efficient, and accurate in their delivery methods, particularly for data communications. Changing American telecommunications policy and regulation, then, would allow new telecom players to compete against AT&T. The ensuing policy shift was a long, complex process with many legislative and regulatory battles that began in the late 1960s and intensified in the 1970s and 1980s. Large telecommunications user groups, research institutes, foundations, universities, and specific federal departments and the FCC all contributed to neo-liberal telecommunications policy and regulatory reform through deregulation and liberalization.

Jill Hills (1986: 50) goes so far as crediting the United States with being the first deregulator. Others, such as Mosco, call for a reassessment of the terms regulation and deregulation. He writes, "the policy debate over deregulation is disingenuous at best, because deregulation is not an alternative. Rather, the debate comes down to the choice among a mix of forms that foreground the market, the state, or interests that lie outside both. Eliminating government regulation is not deregulation but, most likely, expanding market regulation" (1996: 201).

Building on Mosco, a more appropriate term is neo-regulation. This is where the state replaces regulation that was previously based on the public interest and on service, afford-ability, and universality with market regulation. As a broad term, neo-regulation has four characteristics:

1 *No regulation*, where market regulation is decided by a competitive environment that has a number of telecommunications players. Oligopolies (three to six telecommunications players all offering similar products at similar prices) may not be regulated or may claim to be self-regulating.
2 *Managed regulation*, or light regulation of the established telecommunications providers, which is still carried out by the existing state regulators. It may also tend towards state management of competing telecommunications interests, where the state may prevent particular players from competing in particular service areas.
3 *Re-regulation*, where the state has to step in because market-regulation failure or abuse is so noticeable to the public that a portion of the non-regulated area needs regulation again.
4 *Forbearance*, as in the Canadian case, where the state chooses to not regulate what it considers competitive areas. The Canadian regulator, the CRTC, may, however, step back in and re-regulate if market failure occurs.

Liberalization "is a process of state intervention to expand the number of participants in the market, typically by creating, or easing the creation of, competing providers of communication services" (Mosco, 1996: 202). The state changes existing policy and regulation by allowing private competitors into the telecommunications arena, thereby increasing market competition. As the following account of telecommunications policy and regulatory reform in the United States indicates, often the state regulator is engaged in neo-regulation and liberalization policy reform at the same time.

A combination of entry liberalization, no regulation, and managed regulation by the FCC allowed private networks to compete with the public telecommunications network. In the Above 890 microwave decision in the late 1960s, the FCC allowed an alternative to the Bell system – that is, private networks for private use. Using the expanded radio frequency above the 890 frequency, microwave technology would provide transmission through private lines for large business users such as the petroleum, railway, and power companies. An alliance of prospective private users and equipment manufacturers all lobbied the FCC for permission to use the 890-megacycle spectrum. Microwave manufacturers such as the members of the Electronic Industries Association presented an extensive engineering study showing the feasibility of using microwave service through the identified spectrum. A number of large business users, represented by the political agencies of various sectoral organizations such as the American Newspaper Publishers Association (ANPA), Montgomery Ward, the National Retail Dry Goods Association (NRDGA), the Automobile Manufacturers Association (AMA), the National Association of Manufacturers (NAM), and the American Petroleum Institute (API), provided supporting evidence at the hearing that a microwave system would be more economical, faster, would be able to handle larger volumes of data, and would have a higher degree of reliability (Schiller, 1982: 9–12). The business lobby organizations supported a microwave system because it could be tailored to the large users' communication needs.

At the hearings, AT&T led the opposition of the common carriers, arguing that private transmission systems would lead to cream-skimming. In addition, AT&T warned that allowing competition with the public network would eventually harm universal service and lead to increases in the cost of communications services to individuals, small business, and the economy in general.

By 1969 the FCC, in the MCI decision, liberalized entry for Microwave Communications Inc. to construct a microwave link between Chicago and St Louis. Large business users, computer-service organizations, and microwave-equipment manufacturers such as Bethlehem Steel, Union Carbide, Dupont, American Express, Chrysler, Monsanto, API, NAM, NRMA, and the American Bankers Association opposed AT&T's monopoly position, arguing that they needed the bulk transmission capacity that MCI would offer to meet their interoffice and interplant needs (Schiller, 1982: 43). AT&T and the other common carriers argued that competition would adversely affect the entire rate structure and subsidization system (Schiller, 1982: 43; Horwitz, 1989: 227–8). Not only did the FCC grant MCI limited competition; in other decisions from 1971 to 1974 the regulator allowed the interconnection of MCI's long-distance data lines with the public local networks. Because of a former FCC decision, the Computer I Inquiry, the MCI network was not subject to regulation because it was considered a data-communications system.

In a number of subsequent decisions through the 1970s the FCC intensified its regulation of AT&T in order to manage disputes initiated by the monopoly. AT&T denied interconnection and offered TELPAK, a bulk leased service, at rates far below those of its competitors, and well below cost. Regulatory decisions ordered AT&T to interconnect with the private-line competitors. Western Union and Motorola charged that AT&T priced its TELPAC service rates artificially low and that the FCC should discontinue the service. Four years later, the FCC ruled that AT&T was indeed offering this service at predatory rates.

In the mid-1960s the FCC undertook an investigation of the telecommunications and computer industries. An extensive study conducted by Booz, Allen and Hamilton for the Business Equipment Manufacturing Association identified a number of data problem areas that large telecommunications users had with the common-carrier services and networks. Problems ranged from data speed to computer interfacing, the error-proneness of the common-carrier system, and the rate structure of the telecom system. The report identified more than 800 individual computer-data users in a number of sectors, including airlines, banking, computer services, education, manufacturing, and the federal government (Schiller 1982: 32). At the inquiry a number of presentations were made by API, NRMA, NAM, ANPA, the Aerospace Industries of America, Athena Life, the American Bankers Association, the American Trucking Association, and the Association of American Railroads to oppose regulation of computer-data carriers. In its final report (Computer I Inquiry) the FCC created an artificial boundary that separated voice communications from the transmission of data through communications networks. In this important decision, voice communications and the carriage industry would continue to be regulated, while the computer industry and data-processing would not. Large business users were very active in the investigation, expressing their concern over the direction and shape of American telematics (Schiller, 1982: 28–32; Hills, 1986: 64–5). This policy decision had far-reaching regulatory implications throughout the 1970s, as new private networks and business service providers, such as resellers, involved in data transmission benefited by not being regulated.

In the mid-1970s the FCC granted liberalized entry to a number of new telecommunications private-service competitors to provide a number of business services. Most prominent were cellular, enhanced services (also known as value-added network carriers, or VANs) and resale services. Enhanced services offered by Telenet and Tymnet provided business communications services, which were developed from the interaction of telecommunications services with computer hardware, software, and labour. Resellers such as Packet Communication Inc., Graph Net System Inc., and Telenet Communication Corp. engaged in the secondary sale or lease of telecommunications services that were originally purchased, generally at bulk discount rates, from AT&T or the regional Bell companies (Horwitz, 1989: 228, 353). The FCC approved all the previously mentioned applicants, but it did not regulate the resellers or the enhanced-service providers. The commission was of the view that market regulation would prevail in this new open and competitive environment. Citing policy that was established in Computer I, the FCC prevented the common carriers from competing in the reselling and enhanced market. As Schiller (1982: 43) correctly points out, liberalization and neo-regulation were not simply an extension of a new pro-competitive philosophy. Rather, the needs of the heavy corporate telecommunications users for flexible cheaper services took primacy over the communications needs of the general public.

Ten years later, the Computer II Inquiry was conducted by the FCC. Essentially, the outcome of the inquiry was a retreat from traditional public-utility regulation and an entrenchment of neo-liberalism. An underlying issue of the inquiry was how the national telecommunications network, dominated by AT&T, could be upgraded to enable data transmission without more regulatory supervision. Computer II gave AT&T the freedom to engage in network-wide computer communications and the right to compete with the other private resellers and enhanced-service providers. Consequently, AT&T moved aggressively into this unregulated environment. But extensive lobbying by a coalition called the Competitive Alternatives for Users Service and Equipment on Tele-Cause[1] insisted that deregulation and true competition could only be guaranteed if AT&T continued to be subject to regulatory constraints and safeguards. This lobbying resulted in an amendment to Computer II that allowed AT&T to compete in computer communications services but subjected the company to regulation (Schiller, 1982: 90).

Political activism was intense over competition in long-distance services. It took two attempts before MCI was finally allowed to compete in the long-distance telephone-service market after the U.S. Court of Appeal overturned the FCC's decision. An ad hoc Telecommunications Users Committee, made up of American Express, Control Data, Dupont, Ford, General Electric, J.C. Penny, K-Mart, Monsanto, Olin, Reynolds and Reynolds, Sears, U.S. Steel, Visa, Westinghouse, the Aerospace Industry Association, and API, and the Committee of Corporate Telephone Users (CCTU), comprising thirty-five users that included Citicorp, Time Inc., Avis, and Greyhound, made presentations to the FCC for competition in long-distance services (Schiller 1982: 76–7). The consequences of the resulting ruling (*In the Matter of Policy & Rules Concerning Rates for Competitive Common Carrier Services*,　Docket No. 79–252, FCC, 1981) to allow competition in long-distance services tended to make the boundary between regulated and competitive telecom services even more ambiguous.

The Communications Workers of America (CWA) also initially supported and accepted the emerging competitive regulatory environment in areas that affected its membership, such as customer-premise equipment and long-distance services. CWA's pro-competition support was ensured once AT&T promised to alleviate the harshest aspects of company restructuring, which gave the union some control over the effects of technological change and a provision that laid-off workers would get good severance packages or retraining (Winseck 1993: 123). Horwitz explains that the paradox of American telecommunications policy reform occurred because liberal groups and other public-interest advocates took the same position that conservative free-enterprisers took by criticizing the telephone monopolies (1989: 16, 213). Ralph Nader's consumer-advocate organizations condemned price and entry regulation, stating that it was a prime example of regulatory capture. Throughout the 1970s these public-interest groups pressured the state public utility commissions to deny the Bell companies local telephone rate increases. These groups also made presentations at the FCC long-distance and customer-premise hearings, coming out in favour of pro-competition policy (ibid., 213). The public-interest advocates supported liberalization and competition as a solution to entrenched AT&T corporate power. This reveals that the shift in U.S. telecommunications policy to liberalization and neo-regulation came about because most of the major policy actors – the large business users, their lobbyists, the new telecommunications challengers, research organizations, the executive departments of the federal governments, and the FCC, as well as the telecommunications unions and consumer advocates – were on the same side. Only AT&T and the other common carriers opposed competition. In addition, the public-interest advocates of anti-monopoly and a pro-competitive stance helped to fuel the anti-trust case against AT&T.

THE FEDERAL JUSTICE SYSTEM AND THE MODIFICATION OF FINAL JUDGMENT

The first anti-trust case against AT&T culminated in a consent decree in 1956. This decision was beneficial for AT&T because it did not break up the vertically integrated monopoly.[2] It did, however, confine the company to providing only common-carrier services that would be regulated by the FCC. Once the FCC started, in the 1950s through 1970s, to create a competitive environment, AT&T intensified its anti-competitive behaviour by doing things like denying interconnection to the new competitors, or creating competing services and offering them at predatory prices in order to retain or increase market share. By 1974 the Justice Department brought an anti-trust suit against AT&T, and MCI filed a private anti-trust suit against the company as well. The Justice Department alleged that AT&T's integrated telephone network engaged in monopoly activity not only with its telephone equipment but with its long-distance service. The government further charged that, as long as the company controlled the local circuits that provided most consumers with their only means of access, competition would not occur in long-distance service, nor would competition develop in data services, private exchanges, telephone systems, telephone switching, or other telephone services and equipment (Crandall and Flamm, 1991: 16). As Schiller notes, the Justice Department charged AT&T with relying on its position of dominance to keep competitors out of long-distance transmission and equipment manufacturing, while at the same time it maintained its local-franchise monopolies (Schiller, 1982: 91).

Normally, American government policy changes are made by Congress, then carried out by various government agencies, such as the FCC for telecommunications. Derthick and Quirk explain that this sequence was reversed with the deregulation of the transportation and telecommunications industries (1985: 96). Telecommunications policy reform in particular originated with the FCC and the pro-competition large telecom-user companies. Although the government did show support for competitive telecom policies, Congress was less successful in passing a reform bill that would restructure AT&T's monopoly. In 1976 AT&T responded to what it considered government meddling with the proposed Consumer Communications Reform Act (the Bell Bill). If passed in Congress, the bill would have restricted competition in transmission services at the federal level and in terminal equipment at the state level. Despite support by the independent telephone companies, by late 1977 Congress finally rejected the bill (Derthick and Quirk, 1985: 98–9).[3] For political reasons the reform focus in Congress, however, did not have the same consensus and clarity as academic opinions. The advantages of competition in telecommunications were not as clear to Congress as were those regarding transportation reform. There was some agreement on the benefits of competition and interconnection, but competition in long-distance telephone service was a horse of a different colour. Doubts continued because of the so-called subsidy that long-distance revenues provided to rural and residential services (ibid., 100).

Until 1974, AT&T was able to avoid the serious anti-trust flak that would result in any major restructuring of the company. But that year the political will of the Justice Department prevailed in the *United States v. American Telephone and Telegraph Co.* suit, which resulted in the break-up of AT&T ten years later. The suit charged that the company had monopolized interstate communications and the market for telecommunications equipment (Crandall, 1991: 36). According to the Justice Department, the only fair solution to these practices was the divesture of Western Electric and Bell Laboratories from AT&T. After six years of bickering between the Justice Department and AT&T, the case was finally brought to trial before the Federal Court. Once the government case was concluded, AT&T moved for a summary judgment. Judge H. Green denied the motion and concluded that the government had provided ample evidence of serious corporate action, including denial of access, aggressive pricing practices, and the sole use of Western Electric manufacturing equipment, all of which prevented new competition and frustrated independent manufacturers (ibid., 38). The final verdict stated that "the testimony and the documentary evidence adduced by the government demonstrated that the Bell System has violated the antitrust laws in a number of ways over a lengthy period of time" (in Derthick and Quirk, 1985: 200). This decision forced AT&T to settle the case. In an agreed settlement between AT&T and the state, called the Modification of Final Judgment (MFJ-1982), Judge Green presented a compromise decision in which most of the vertical structure of the company, its long-distance operations, manufacturing and research and development subsidiaries were left intact. AT&T, did, however, have to divest itself of its twenty-two Bell operating companies. Seven new independent regional Bell operating companies (RBOCs) were created, but they were still subject to PUC regulation because of their monopoly position vis-à-vis basic telecom service.

One of the major differences between the Canadian and American governments' roles in the restructuring of their respective telecommunications companies is that the Canadian government did little more than approve BCE's reorganization plan, whereas the U.S. federal court system played a key role in breaking up AT&T, which helped to create an even more powerful telecom sector.

The settlement was a victory for AT&T, the RBOCs, the large users, and the new competitors. In the very short term, as a result of divesture, AT&T's market share dropped to 65 per cent of the long-distance market. By 1988 the new competitors, MCI and Sprint, held 25 per cent of the long-distance market share. None the less, AT&T's earnings recovered and indeed increased through the 1980s and 1990s as it entered the unregulated service area and international market (Sherman Report, 1988: 34–5). Not only did the large users benefit from lower long-distance rates; the RBOCs expanded internationally by purchasing shares in former PTOs that had been privatized, as well as other communications businesses. This was at the expense of U.S. national telecom capital expansion, as well as a cause of a major increase in quality of service problems.

But the policy changes of liberalization and neo-regulation did not bring about a win-win situation for everyone. The largest losers were the residential consumer groups and the telecommunications unions. After divesture, from 1984 to 1989 local rates jumped 43 per cent, while interstate toll rates and long-distance rates dropped 5 per cent and 35 per cent respectively (Mosco, 1990a: 14). The very poor in the United States could only stay on the telephone system or gain access to it through two newly created telephone welfare progams, Link-up America and Lifeline (Aufderhide, 1987: 91–4; Mosco, 1991: 2–4).

Although AT&T and the telecom unions had presented a joint political campaign to prevent the break-up of the Bell system, after divesture, labour-management relations abruptly changed. First, national bargaining was eliminated by the newly created RBOCs and replaced by corporate and subsidiary-level bargaining. This resulted in the second-largest strike in United States history, when 600,000 hourly CWA members walked off the job in 1983. In 1986 the CWA struck AT&T over wage reductions and the elimination of the cost-of-living allowance (COLA). Other strikes took place over the elimination of health benefits and the introduction of managed health care. In addition, total telephone employment had dropped from 1982 to 1992 by 222,000 employees. Employment in the RBOCs had declined by 14 per cent, with a loss of 158,000 people. Employment also dropped at AT&T from 1984 to 1992 to 124,300, a decline of 33 per cent of its work force. Prior to divesture some 80 per cent of all telecommunications employees were in unions, but by 1988 this figure had dropped to 70 per cent. The new telecom service providers, by contrast, are noted for their anti-union stand. For example, Sprint's level of unionization is only 12 per cent, and MCI's is 25 per cent (CWA, 1994, Preserving High-Wage Employment in Telecommunications, 2, 4, 7; ILO, 1993: 132). With the extension of liberalization and competition in telecommunications into local services, the RBOCs and AT&T have continued to downsize, eliminate jobs, ignore service quality, increase local telecom rates, reduce long-distance rates, and develop the non-regulated service areas, while at the same time continuing to benefit from scoring revenues throughout the second half of the 1980s to the 1990s.

EXPORTING THE U.S. NEO-LIBERAL TELECOMMUNICATIONS POLICY MODEL

Multinationals intensified their political pressure on the U.S. government to have the neo-liberal telecommunications-policy model adopted in other countries. The multinationals wanted free entry and free access for their goods and services in other national economies. Breaking down and breaking up the protective policies of other countries was necessary so that these businesses could integrate their world-wide activities in manufacturing, marketing, shipping, finance, insurance, communications, and advertising. It required the U.S. federal state to get actively involved in co-ordinating and exporting its reorganized telecommunications policies to other countries, and influencing these countries to accept the free-market philosophy.

The U.S. federal government played a key role in addressing the demands of the multinationals by conducting hearings in the House of Representatives that culminated with the passing of the International Communications Reorganization Act of 1981. American Express testified, as did a number of other multinationals, that protectionist barriers in other countries needed to be broken down. In a detailed table Dan Schiller identified trade barriers to data, telecommunications, and information services on a country-by-country basis (1982: 124–35). According to the table, what was needed was policy liberalization for enhanced or value-added services, and private networks or user-controlled networks, as well as an "open-sky" approach to international satellite policy.

In 1970 the U.S. government moved to co-ordinate its domestic and international telecommunications policy through its executive branch by establishing the White House Office of Telecommunications Policy. The office was moved to the Department of Commerce in 1978 and renamed the National Telecommunications and Information Administration (NTIA). The Office of Plans and Policy also co-ordinated telecommunications policy within the FCC (Mosco, 1989). Within the executive, the Council of International Communications and Information was created with the specific mandate to co-ordinate the policies and activities of all federal agencies involving telecommunications and information, and to review all policy determinations of federal agencies and all proposed statements of U.S. policy by such agencies relating to international communications and information, and approve, disapprove, or modify any such policy determination or proposed statement (cited in Schiller, 1982: 141). Other executive offices, such as the secretaries of State and of Commerce, the director of the Office of Management and Budget, and the assistant to the president for National Security Affairs, as well as the FCC, not only co-ordinate U.S. telecommunications policy but integrate it into the nation's trade policy (ibid.).

Changing the hegemonic project to reflect neo-liberal policy reform in the United States also brought about very significant changes in social relations. Liberalization and neo-regulation replaced telecommunications universality in the public interest with a model that served market interests and was backed by market regulation. Policy changes resulted in large benefits for the large business users, predominately multinationals and conglomerates, the new service and system providers, and the established providers. The enormous cost of policy restructuring, however, was borne by residential and small-business subscribers, those who lived in rural areas, the poor, and the telecommunications workers as their local rates soared, accompanied by poorer service. Divesture and competition led to lay-offs in the industry, particularly a reduction in the unionized work force. In addition, dramatic changes occurred in labour relations over bargaining, wages, and benefits.

These changes in telecommunications social relations are important to keep in mind when we examine the subsequent policy shift that began in Canada in the 1970s. As the neo-liberal telecom policy model was applied in Canada, it produced policies, legislation, and regulation that would integrate telecommunications policy into the North American continental orbit. Even though the United States went to great lengths to export their telecommunications policies, the policy shift in Canada occurred because of a complex set of external and internal social forces promoting telecom neo-liberalism, which was none the less strongly resisted by the majority of the population.

INTERNAL CANADIAN FORCES

Telecommunications and Canada: Consultative Committee on the ImplicationsofTelecommunicationsforCanadianSovereignty (Canada 1979), also known as the Clyne Report, conducted for the minister of Communications, then Jeanne Sauvé, reveals the conflicting and contradictory position the federal government found itself in regarding telecommunications policy. The Clyne Report was very concerned with external (read U.S.) market forces and competitive telecommunications policies. According to the report, the federal government needed to exercise its sovereignty[4] over telecommunications. This included providing adequate measures of control over data banks, transborder data flows, and regulating information-service content (ibid., 84). Independence over the Canadian-owned telecommunications sector could be realized with the continuance of the mixed telecommunications system for the satellite and telephone industries. The Clyne Report was, however, concerned with the briefs and submissions from other federal government departments, large user organizations, and new service providers, such as the Canadian Manufacturers' Association, the Canadian Business Equipment Manufacturers' Association, the Canadian Association of Data Processing Service Association, Canadian National and Canadian Pacific Telecommunications, IBM, and the Economic Council of Canada (96–7). These users and new telecommunications challengers were unanimous that long-distance telephone and business service rates were too high, and that telecommunications policy covering these areas needed to be reviewed. What was so ironic about the report was its concern that a heavy price would have to be paid for competition. The report was concerned about upsetting the existing pattern of cross-subsidization, which kept local residential telephone rates affordable for all Canadian regardless of where they lived: "telecommunications, as the foundation of the future of society, cannot always be left to the vagaries of the market; principles that we might care to assert in other fields, such as totally free competition, may not be applicable in this crucial sphere. We must look at it freshly, without preconceived ideas" (2).

But, as the Clyne Report recommended strengthening Canadian telecommunications sovereignty, another federal institution, the CRTC, through a series of public hearings, introduced telecommunications policy liberalization and neo-regulation, paving the way for a competitive telecommunications environment. Closer scrutiny of these hearings and the commission's final decisions on interconnection, terminal attachment, business services (cellular, enhanced, resale, and sharing) shows that the winds of change for continental neo-liberal telecommunications policy came from forces within Canada as well as without. The major difference between Canada and the U.S. was that the Canadian telecom policy changes were fiercely resisted by the established telephone carriers, the unions, public-interest advocates, consumer groups, and many of the provincial governments. The telecommunications unions, the public-interest advocates, and the consumers' organizations in particular were responsible for stopping the first attempt by CNCP to compete in the long-distance telephone market. Although dominant carriers such as Bell Canada initially resisted policy liberalization, to avoid the continued regulation of all of its activities and to avoid the break-up of its vertically integrated operations, Bell reorganized itself into Bell Canada Enterprises and at the same time embraced the free-trade aspects of neo-liberalism.

LIBERALIZATION AND NEO-REGULATION

The first major liberalization hearing took place in 1979, after CNCP applied to the CRTC to interconnect its telegraph network and its private voice and data networks to Bell Canada's public switched telephone network (PSTN).[5] The CNCP Telecommunications: Interconnection with Bell Canada hearing was the largest public hearing up to that time. It included two pre-hearing conferences, twenty-five days of public hearings, numerous submissions, testimony by expert witnesses (economists), as well as telecom business lawyers, and consultants (Telecom Decision CRTC 79–11, 1979; see Appendix 9 for a complete list of participants).

CNCP Telecommunications' main argument was that Bell's refusal to allow CNCP to interconnect to its network was an abuse of its monopolistic position. In CNCP's view, interconnection would allow the company to compete better with Bell in providing private data- and voice-line networks and services (CRTC 79–11, 68). As Table 4.3 reveals, CNCP provided evidence showing the asymmetrical distribution of Canadian telecommunications revenues. CNCP presented itself as a David battling the Goliath Bell over the latter's refusal of interconnection, restricting CNCP to only 5 per cent of the telecommunications market.

At the time of the hearing CNCP's private data and voice networks were custom-designed to meet the special requirements of its subscribers, including the air-traffic-control system for Transport Canada, the federal government's meteorological network, the inter-police network for the RCMP (Ottawa), and a private data-line network for the Social Affairs Department of the Quebec government (CRTC 79–11, 13). According to CNCP, interconnection liberalization would provide significant benefits to other telecommunications users, giving them choice in telecommunications facilities and services.

A number of telecommunications-user organizations, data-processing and computer industries, two provincial governments, and the federal government all supported CNCP's application. User organizations such as the Canadian Manufacturers' Association (CMA) and the Canadian Business Equipment Manufacturers' Association (CBEMA) stated that interconnection would lead to increased competition in private voice and data networks, which in turn would improve efficiency and reduce telecommunications costs. The Canadian Industrial Communications Assembly (CICA), representing 130 of the major telecommunications users, supported CNCP's application so that its members could have greater freedom of choice in selecting telecommunications services. The Canadian Information Processing Society (CIPS), representing 3,000 computer professionals, also supported more freedom in choice of telecommunications networks and services. A number of telecom users such as the Canadian Press, IBM Canada, the Business Interveners Society of Alberta, the Royal Bank, the Bank of Nova Scotia, Interprovincial Pipelines Ltd., Westinghouse Canada, General Electric, and Domtar supported CNCP's request to interconnect to the Bell network. The Canadian Petroleum Association (CPA) even went as far as supporting complete interconnection to Bell's network and to all the other telephone carriers (CRTC 79–11, 70–1).

The federal government's Director of Investigation and Research, Combines Investigation Act (here after, the Director, now the Director of the Competition Act & Competition Bureau) and the provincial governments of British Columbia and Ontario were also strong supporters of interconnection liberalization. All three government participants held views similar to those of the large users. Interconnection would lead to competition in the communications industry, improve

Table 4.3
Distribution of CNCP, Bell Canada, and other telco revenues by market share
and segment as of 1976 ($ millions)

Carriers	Data computer	Data message	Voice private line	Voice public telephone	Voice broadcast	Voice other	Voice total
Bell	80	9	84	1,674	11	46	1,904
All other telcos	133	17	113	3,005	22	74	3,364
CNCP	17	139	8	24	4	0	192
Telco % share	89	11	93	99	14	100	95
CNCP % share	11	89	7	1	8	0	5

Source: Adapted with permission from CRTC Decision 79–11, 1979, p. 18.

communications efficiency, and provide more consumer choice. But both provincial governments wanted regulation to continue (CRTC 79–11, 71–2).

Bell Canada, the provincial telephone companies, and the TCTS opposed CNCP's interconnection request because they considered the application a way of using local telephone facilities to capture a portion of the long-distance market and revenues. In the TCTS members' view, the adverse effects of interconnection would result in increases in local telephone rates, a 27 per cent increase for residential subscribers, and a 37 per cent increase for business subscribers (CRTC 79–11, 66). Taking a very arrogant position, Bell warned the CRTC that long-distance revenues are "rightly due Bell," and threatened that if its monopoly position were to change, the existing arrangements of basic universal telephone service for the public would be undermined (66).

SaskTel, by contrast, opposed interconnection liberalization because its potential benefits would be confined to narrow interests (telecom users and the new telecom challengers), jeopardizing the telecommunications infrastructure for all Canadians at the lowest cost. SaskTel even provided evidence and witnesses regarding the American experience with interconnection liberalization, which had resulted in higher costs, duplicate facilities, increased rates for households and small businesses, and numerous competition violations (CRTC 79–11, 74). AGT pointed out that interconnection liberalization was not in the public interest as it would only benefit the private interests of CNCP (76). The Canadian Federation of Communication Workers (CFCW) opposed liberalization because it considered interconnection the slippery side of competition, which would threaten existing job stability in the industry (74). The government of Quebec also opposed liberalization, for two reasons: it would benefit private, not public interests, and the CRTC would be passing judgment on communications matters that were important to Quebec sovereignty. Similarly, the Atlantic provincial governments opposed interconnection liberalization, calling instead for a federal-provincial task force to study the entire question of communications competition (79).

But despite the opposition of the telephone companies, most of the provincial governments, and the telephone unions' federation, the CRTC approved CNCP's application. The CRTC concluded that "competition would be greatly enhanced with interconnection" and that it was "in the public interest" (242). This decision is of prime importance in the history of the liberalization of telecommunications policy because the CRTC explains what it meant by the "public interest," which affected all subsequent liberalization decisions. The public interest was defined as "important for users of communication services in Canada" (240).

At the same time the CRTC continued its practice of managed regulation of CNCP and Bell Canada by preventing CNCP's interconnection to Bell's public long-distance network, overseeing CNCP's compensation to Bell, as well as keeping track of the evolving nature of the competitive market. Subsequent interconnection hearings and decisions allowed CNCP to interconnect to BC Telephone's local network; BC Rail to interconnect to BC Telephone; and the interconnection of intra-exchange systems. Although the applications were opposed by the telephone companies and the CFCW, the support from the Director and the large user organizations influenced the CRTC more, as reflected in its decision to approve both applications (CRTC 81–24, 1981, and 85–19, 1985).

BUSINESS SERVICE LIBERALIZATION

Further liberalization of entry occurred when the CRTC approved a number of business services. The first was the attachment of subscriber-provided terminal equipment to that of the Bell, BC Telephone, and CNCP networks. The Ontario Hospital Association (OHA) and the Telephone Answering Service of Canada (TASC), in particular, wanted to be able to connect their telephone equipment to the existing networks. Bell and BC Telephone argued that Rule 9 of Section 321 of the Railway Act gave them the exclusive right to require customers to use only their equipment. According to the OHA and TASC, this gave unreasonable advantage to the carriers. At the terminal-equipment hearing the familiar pro-liberalization policy supporters, including the CMA, the CBEMA, the CPA, and the Retail Council of Canada, supported telephone-equipment and -attachment liberalization because it would usher in new suppliers and open new markets for them (CRTC 82–14, 1982: 28). State support for terminal-equipment and -attachment liberalization came from the Director and the provincial governments of British Columbia and Ontario. Unlike its position in the interconnection hearing, the Consumers' Association of Canada (CAC) was in favour of terminal-equipment liberalization because it would eliminate the total control that the telephone companies had over network development. In addition, the CAC was of the view that consumer needs would be better served by competition in telephone equipment (CRTC 80–13, 81–19, 82–14).

In its conclusion the CRTC approved liberalization entry of terminal equipment and terminal attachment (CRTC 82–14). The commission was not convinced of the potential disadvantages of liberalization entry raised by Bell and BC Telephone, or of the invasion of foreign manufacturers. This would suggest that the CRTC was well aware of the changes taking place with the internationalization of the market. Nor did the CRTC seem concerned by the CFCW's evidence that terminal-attachment liberalization would result in the loss of end-to-end control of the networks, or TWU's claim that terminal equipment liberalization would have harmful effects on installation and repair jobs (ibid., 25).

Referring to statements made by the various pro-liberalization supporters from user organizations, government representatives, and the CAC, the CRTC stated that the disadvantages were not significant enough. In the CRTC's view, the benefits of enhanced consumer choice, lower prices through competition, and increased flexibility and efficiencies for business subscribers were far more important (CRTC 82–14, 1982: 29). Nevertheless, the CRTC continued its approach of managed regulation by establishing technical standards and terms and conditions of attachment (ibid., 31–2).

Liberalization of entry continued in business services when the minister of Communications (then Francis Fox), through DOC, issued a non-telecommunications licence for a new telecom-delivery system: mobile telephone radio communication or cellular radio telephone. This technology supposedly made more efficient use of scarce radio frequency. Cellular communication was considered useful for businesses and professionals. One of the major power distinctions between DOC – now Industry Canada – and the CRTC is that the CRTC sets rates and the terms of interconnection, whereas DOC has the power to issue spectrum licences. Essentially, DOC created a duopoly in cellular phone services with its awarding of the spectrum licences to

Cantel (major shareholders included CNCP, Rogers Communications, and Motorola Canada Ltd.) and the telephone-carrier cellular operators, Bell and the provincial telephone companies (Surtees, 1993: 121).

Support came from the Director, the governments of British Columbia and Ontario, the large telecom-user lobby organizations, and equipment manufacturers such as the CICA, the CMA, the CPA, and the Electrical and Electronic Manufacturers' Association. Bell and BC Telephone were opposed to the radio common carriers' interconnection to their public network because they already offered cellular services. BC Telephone and NorthwesTel and Terra Nova were opposed to Cantel's entry because it would lead to market fragmentation, the erosion of revenues, and toll (long-distance) bypass (CRTC, 84–10, 1984: 10). As in the previous pro-liberalization entry rulings, the CRTC stated that a cellular duopoly would provide users with more choice, lower prices, and benefits from the new technology. And, as in the other business-service liberalization policies, the CRTC continued its practice of managed regulation of the Bell and BC Telephone cellular-service offerings. The commission called for structural separation of both carriers in order to establish a separate cellular subsidiary so that they would compete fairly and without cross-subsidization of revenues from their public systems. Later in 1984 the federal government designated Cantel Inc. the national cellular-radio-telephone service provider. In addition, Cantel was not required to file tariffs and would not be regulated by the CRTC. Once Bell and BC Telephone established separate cellular divisions, they too were not subject to CRTC regulation, so that cellular services would evolve "in an unregulated, and as competitive environment as possible" (CRTC Telecom Public Notice 1984–55, 1984c: 1).

A series of enhanced-service CRTC hearings took place throughout the 1980s to decide entry liberalization. Enhanced service combines computer-processing and telecommunications technology, which is then used to create a new range of telecommunications services, combining information processing and storage. In two hearings the CRTC totally liberalized entry for all enhanced-service providers (CRTC 84–18, 85a–17). The continental implications of the enhanced-service decisions became apparent when the CRTC issued a public notice for the upcoming hearing and recommended that Canada adopt the FCC definition of basic and enhanced-services (CRTC Telecom Public Notice 1983–72). At the enhanced service hearing, state participants such as the Director and the government of Quebec, as well as the most significant user organizations, including the CICA, CIPG, and the Canadian Industrial Communication Assembly et al,[6] all supported adopting the American definition (CRTC 84–18, 1984: 6–12). The government of Ontario and Telesat Canada objected to the outright adoption of the U.S. definition, stating that there were major differences between the Canadian and American industry structures. The CRTC modified both definitions somewhat to read:

a) basic service is one that is limited to the offering of transmission capacity for the movement of information;
b) an enhanced service is any offering over the telecommunication network which is more than a basic service.   (CRTC 84–18, 1984: 12, 14)

State actors such as the CRTC, the Director, the government of Quebec, and the political agents for the large telecommunications-user organizations for all intents and purposes adopted what Coulter calls a "continental standard" for enhanced and basic services (1992: 241). Adopting this continental standard would become even more significant once trade in telecommunications services was written into the continental trade agreements, the Canada-U.S. Free Trade Agreement (CUSFTA) and the North American Free Trade Agreement (NAFTA).

CUSFTA is important because it is the first trade document to extend free trade beyond goods to services. Pertinent telecom services include those that are enhanced or add value to computerized data, audio, video, and information services. A whole chapter is devoted in NAFTA (chapter 13) to telecom services. Enhanced services are also crucial to the Internet and the World Wide Web for multimedia (new media) content and products. Multimedia content is a value-added service because it employs computer-stored or restructured processing applications that include formatting, content, codes, and even the protocols of transmitted information.

Essentially NAFTA creates a free-flowing data and information zone that permits business to move information freely throughout the North American continent. To date in Canada and the United States, multimedia is not subject to government regulation. At the international level the World Trade Organization (WTO) and NAFTA act as vehicles to promote liberalization, deregulation, and internationalization. In essence, the WTO and NAFTA establish non-transparent trade for multinational firms and for international communications-service users.

Major domestic opposition occurred in the United States, Canada and Mexico over the undemocratic aspects of NAFTA. The emphasis in these trade agreements is economic relations and the advancement of the "free market." Social or public policy concerns, by contrast, are absent, as are citizens' democratic rights or public-service responsibilities (Rideout and Mosco, 1997: 99).

In its final comments the CRTC stated that "the principles of competition should guide the enhanced service market" (CRTC 84–18,

1984: 18). The only opposing voice came from the CFCW, who argued that competition in business services would come about at the expense of ordinary consumers. The CRTC disagreed with the minority view offered by the CFCW. The commission liberalized entry for enhanced-service providers and decided not to regulate the non-common carriers. The common carriers were allowed to compete with the non-common carriers, but they would continue to be regulated by the CRTC. Most of the previously mentioned user groups, the Director, and the government of Quebec agreed that if there was no structural separation of the carriers' enhanced services, then they needed to be regulated to avoid cross-subsidization (CRTC 84–18: 45–6).

Policy liberalization of enhanced services led to the intensification of class disputes and conflicts between the new enhanced-service providers and the telephone carriers. In another hearing, Bell and CNCP applied to the CRTC to deny the resale of Call-Net Communications Ltd. (Call-Net) enhanced services. According to Bell, Call-Net's services were not enhanced, only basic. The CRTC agreed (CRTC 87c–5). Call-Net fought back by requesting the CRTC to review its decision. After the CRTC again decided that most of Call-Net's services were basic, Call-Net's next move was to apply to the federal government to overturn the CRTC's decision. Although the decision was not overturned, a federal cabinet order-in-council extended Call-Net's access to Bell and BC Telephone for one year. These events were repeated when Bell, CNCP, and the CRTC all agreed that Call-Net was illegally reselling a large portion of its enhanced services. A second order-in-council once again extended Call-Net's interconnection to CNCP and Bell (Rideout, 1993: 40–1).

This episode reveals that the process of bringing about telecom policy neo-liberalization in business services helped to intensify class conflicts. Neo-liberal policy changes were intended to liberalize entry so that new telecommunications-service providers and new delivery systems would be able to compete with the established telephone carriers, thereby ending the position of the incumbent telecom monopolies. But these neo-liberal policy changes also had unintended consequences. They brought into question the supposed arm's-length regulation of the CRTC. Consequently, liberalization and neo-regulation increased political interference. It would take until the 1990s and many more hearings before the enhanced-service issue was resolved and the CRTC stopped regulating that particular telecom service.

In the Interexchange Competition and Related Issues hearing, the CRTC began the first of many hearings on liberalizing entry to create a competitive resale and sharing market. Under the Railway Act, resale and sharing were only permitted with the approval of the telecom carriers. The Canadian Business Telecommunication Alliance (CBTA), the Canadian Telecommunications Group (CTG), the Director, and the government of Ontario all lobbied the CRTC to reduce or eliminate these barriers. Opposed were Bell, the provincial telephone companies, CFCW, TWU, and, surprisingly, CNCP. Although CNCP supported policy liberalization in other areas, including long-distance service, it argued that resellers should not be able to acquire facilities from them and then turn around and sell them below cost.

Despite this strong resistance, the CRTC embarked on limited resale and sharing liberalization. Resellers were only permitted into the non-MTS/WATS (message toll service/wide-area telephone service) interexchange services (data only) (CRTC 85–19, 1985: 87). In addition, the CRTC would refrain from regulating this service area, relying instead on market regulation (ibid.). Once resale was liberalized, the resale market doubled every year from 1985 to 1992 (Surtees, 1993: 305–6). By the 1990s more than sixty resellers had entered the Canadian market. Call-Net led the pack, with 35 per cent of the market. Fonorola ranked second, with 10 per cent, followed by the Canadian subsidiary of a British telecom company, Cable and Wireless, with 8 per cent (ibid.).

REORGANIZING BELL CANADA

Bell Canada's response to policy liberalization and the erosion of its telecom monopoly position was swift and aggressive. Unlike AT&T, Bell Canada began reorganizing itself. In the 1960s Bell was successful in having its charter changed so that it could engage in telecommunications, not just the telephone business. Further revisions were made in the 1970s to accommodate a corporation that was expanding internationally, especially though its manufacturing subsidiary, Northern Telecom (now Nortel Inc.), into the U.S. market. It will be recalled that domestic regulatory restrictions prohibited Bell and its subsidiaries from applying for and holding broadcasting licences and cable-television licences (Babe, 1990: 185–6).

By the 1980s, motivated by the emerging trend towards competition and the desire to have fewer of its operations under the watchful eye of the CRTC, Bell filed articles of continuance with the Department of Consumer and Corporate Affairs pursuant to the Canadian Business Corporations Act. In 1982 Bell announced a reorganization whereby Bell would become a subsidiary of its new parent, Bell Canada Enterprises Inc. (BCE). If approved under the Business Corporations Act by Cabinet and the CRTC, Bell would set itself up as a corporate entity rather than continue operating as a public utility. Through restructuring, it would also be able to separate its holdings and move most of its revenues from the scrutiny of CRTC regulation. Prior to the reorganizing scheme, all of Bell's income from both its regulated and non-regulated companies had been subject to CRTC regulation before it reached the shareholders (Coulter, 1992: 389; Babe, 1990: 190). BCE reorganization, then, would reduce CRTC regulation over most of the organization's activities except for its public telephone-carrier operations (CRTC 83–10, 1983: 4).

User groups such as the CICA and the Canadian Telecommunications Group (CTG) supported Bell's reorganization, providing statutory limitations were implemented to prevent rate increases. Also, the CRTC was to continue to regulate BCE to ensure there was no cross-subsidization from its monopoly services (long-distance and local phone revenues) to competitive services (CRTC 83–10, 1983). The Director concurred with the user organizations, but in its submission to the Restructure Trade Practices Commission argued that the public interest would be better served if Bell's vertically integrated system were undone, untying Bell from its manufacturing operations. Other government agents such as the minister of Communication (then Francis Fox) wanted to make sure that the government continued to support increased competition in the high-technology sector. But the minister also stated that Canadian companies needed to be able to make structural changes so that they could be more competitive. According to the minister and the DOC bureaucracy, vertical disintegration was not an option. What was needed for a competitive global marketplace was strong telecommunications players such as BCE and Northern Telecom.

Dissenting voices from the Consumers' Association of Canada (CAC), the National Anti-Poverty Organization (NAPO), and the Public Interest Advocacy Centre (PIAC) held that the reorganization of Bell would have a negative impact on public oversight and universality (CRTC 83–10, 1983). None the less, the CRTC not only approved BCE's reorganization plan but also championed the new corporation. The CRTC's final comments reveal much about the government's neo-liberal philosophy: "Managerial flexibility is particularly necessary at this stage of time. As Canada proceeds into the *information age* its future as an industrial state would depend on high quality managerial, technical, and research skills such as those found within the Bell group of companies" (ibid., 2).

Coulter (1992: 411–n12) correctly ascertained that the reorganization of Bell into BCE was a quick response to changing times. Not only did the reorganization release BCE from public accountability and regulation; it also strengthened BCE's dominant economic position. After reorganization, from 1985 to 1995 BCE ranked first in revenues among all Canadian companies, and through the same period it ranked in the top ten in terms of profits and assets. Even its subsidiaries, Northern Telecom and Bell Canada, ranked in the top twenty-five Canadian companies in terms of revenues (Globe and Mail, Report on Business, 1985 to 1995).

BCE embraced the neo-liberal policy shift in other ways as well. Not only did Bell submit its so-called rate-rebalancing proposal to the CRTC when CNCP applied to compete in the long-distance-service market; it presented rate-rebalancing to the Macdonald Commission. In its brief to the Macdonald commission, Bell submitted that it cost the utility $1.93 to produce $1.00 of revenue from local services. According to Bell, this had resulted in a $1.2 billion cross-subsidy in the previous year, 1982. As Bell stated at the Macdonald Commission hearing, "in time, the pressure to reduce long distance service prices in Canada will become irresistible. Therefore, it is important to Canada to start gradually moving both long distance and local telephone service prices closer to costs" (Bell Canada, 1983: 20).

Rate-rebalancing would have two effects. First, monthly rates for local services would rise, while long-distance rates would fall dramatically. According to Bell, their rate-rebalanced schemes would not generate any extra revenue for the company. BCE (Bell) was aware of the pressures for neo-liberal policy changes both from the United States and from large telecom-user organizations. Its rate-rebalancing plan was aimed at placating the demands of business users for cheaper telecom services while at the same time maintaining its monopoly position as the telecommunications service provider. If Bell could lower its long-distance rates enough, the resultant erosion of profit margins would make entry for any other competitor unprofitable. Bell subsequently submitted to the CRTC its rate-rebalancing proposal along with an alternative to its local flat-rate billing system. The company advocated for a local measured-service (LMS) billing system, a pay-by-the-minute service for local calls, similar to local European telephone service and cellular service. Both proposals, rate-rebalancing and local measured service, were harbingers of things to come.

5 Telecommunications Policy Liberalization and Centralization

In order to compete internationally, Canada will be pushed by the
U.S. situation to move toward cost-based rates, but at its own pace.
This will require gradual decreases in cross-subsidization and therefore
increased local rates. This will hit residential and small business users
hardest, while big business users will benefit.

Department of Communication,
Cabinet Document, May 1985

Supported by the large user organizations and some federal and provincial state agents, CNCP Telecommunications wanted to enter and compete in the public long-distance-service market. In the fall of 1983 CNCP filed an application with the CRTC requesting that it be granted permission to interconnect its system to Bell and BC Telephone so that it could offer message toll service (MTS – basic long-distance service) and wide-area telephone service (WATS – business long-distance packages).

A number of regional hearings and a central public hearing followed one year later, in the fall of 1984. The CRTC rendered its decision, *Interexchange Competition and Related Issues*, in August 1985. Over thirty-six days, 194 exhibits and submissions were filed and approximately 1,700 interrogatories and responses were entered into evidence. Testimony from some of the Telecom Canada members, the challenger CNCP, other communications companies, an array of user organizations, provincial and federal government representatives, telecommunications unions, and consumer and public-interest advocates either supported or resisted the application (for a complete list of participants and presiding CRTC commissioners, see Appendix 10).

In its bid to enter the public long-distance telephone marketplace, CNCP Telecommunications stated that a number of developments had influenced the corporation. They included technological developments that eroded the separate network, and a more general move in Canadian public policy towards competitive private and public telecommunications services (Telecom Decision CRTC 85b–19, 1985: 13). CNCP's major argument for entry into the long-distance market was that the

telephone industry was not a natural monopoly. Expert witnesses for
CNCP explained that if the company were allowed to compete, a
number of benefits would accrue to users, including a reduction in
MTS/WATS rates, greater customer choice and supplier responsiveness,
and an accelerated rate of diffusion of new technology (ibid., 29).
CNCP went so far as to claim that the lowering of long-distance rates
would have a positive impact on Canadian society as a whole (ibid.:
43–4).

CNCP's proposal priced its long-distance rates 30 per cent below Bell
and BC Telephone's rates. The proposal also recommended that CNCP's
level of contribution to interconnect to both companies be 11.5 cents
per minute (ibid., 27, 45). To meet the universal-service requirements
under the Railway Act, CNCP promised the CRTC that it would service
cities with a population greater than 25,000 after its fifth year of
operation, providing it was economically viable. CNCP argued that it
would be able to offer complete universal service in Bell and BC Telephone territories in ten years (ibid., 14). At the end of the initial ten
years, CNCP projected it would obtain approximately 80 percent of
the long-distance market.[1]

It is important at this juncture to recall other statements that CNCP
Telecommunications made regarding universal service. In its submission to the Macdonald Commission, CNCP emphatically stated that
universal service and affordability should not stand in the way of
progress. The following statement is telling: "Public policy in the past
has focussed on keeping the price of access to basic telecommunications service (local telephone service) as low as possible, in order to
achieve universal service ... But we believe strongly that the pursuit of
this goal must not be allowed to restrict unduly the flexibility or choice
of the new, often customized, applications of telecommunications
media" (CNCP, 1983: 4). Clearly this statement reveals that CNCP was
more concerned with its existing and new business customers than it
was with any broad notion of universality and affordability.

Support for CNCP's application came from a number of user organizations in finance, manufacturing, resources, the mass media, and the
service sector, as well as the Director of Combines Investigation, the
governments of British Columbia and Ontario, and the major telecommunications user organizations. The Canadian Business Telecommunication Alliance (CBTA) supported the application because there would
be more customer choice in terms of service providers (CRTC 85b–19:
25). The Ontario Hotel and Motel Association and the CBTA stressed
that new long-distance competitors would provide a variety of service
options. Many user organizations such as the Association of Telecommunication Suppliers, the Canadian Bankers Association, the Canadian

Business Equipment Manufacturers Association, the CBTA, the Canadian Petroleum Association, the Canadian Press, and Cantel expected long-distance competition would provide them with competitive innovation advantages, more choices in new telecom services, and lower rates (ibid., 29).

The strongest federal government supporter for general entry liberalization and long-distance competition was the Director of Investigation and Research, Combines Investigation (the Director). Using only economic criteria, the Director supported the expansion of the telecommunications market. The Director's advocacy for competition in long-distance telephone services was based on the idea that a cornucopia of choice in equipment, services, new products, and lower costs would lead to the creation of more jobs (ibid., 35). The governments of British Columbia and Ontario as well as the Director held the view that aggregate economic benefits would accrue to the whole Canadian economy as a result of allowing competition in public long-distance service. But the government of British Columbia was adamant that the federal government, not the CRTC, should be responsible for changing important national telecommunications policy such as long-distance telephone policies (ibid., 42).

What is evident from the submissions of the various pro-competition supporters is their unwavering ideological belief in the ostensible benefits of an open telecommunications marketplace, despite that fact that CNCP's application lacked empirical evidence and was largely based on weak forecasts, projections, and promises. In fact, CNCP's application was only a seven-page document that did not contain the usual public-opinion polls or marketing evidence to support the bid. One of the reporters covering the hearing commented that the long-distance case was perceived inside CNCP as only an extension of its previous successful interconnection case (Surtees, 1994: 113). Consequently, CNCP treated its application as a relatively simple technical matter – at best, a precise legal issue wrapped in the promise of a brave new telecommunications world.

The supposed merits of the CNCP application were questioned by the Telecom Canada members, the CRTC, and strongly disputed by other provincial governments, consumer and public-interest groups, and the telecommunications unions. Bell Canada and BC Telephone pointed out at the public hearing that CNCP's 30 per cent discount would only be available to CNCP customers. Such discounts would not necessarily apply to the Bell or BC Telephone subscribers, as the large users and the Director expected (CRTC 85b–19, 1985: 27). Participating Telecom Canada members, including Alberta Government Telephone (AGT), BC Telephone, Bell Canada, Maritime Telephone & Telegraph,

New Brunswick Telephone, and Newfoundland Telephone all argued that CNCP would not be able to offer these discounted long-distance rates and at the same time make adequate contribution payments to Bell and BC Telephone.

Bell fought against CNCP's entry into the long-distance market on two fronts: contribution charges and rate restructuring. According to Bell, the existing MTS/WATS rates incorporated an existing average contribution rate of 30 cents per minute. This was 19 cents more than what CNCP was prepared to pay (ibid., 45). In its application CNCP claimed that local rates would not have to rise because it was prepared to make up the lost subsidy from long-distance to local rates through its contribution to both companies, which amounted to 11.5 cents per minute. It is important to point out, however, that this revenue subsidy is based on contentious assumptions and accounting practices that were established almost one hundred years ago, it will be recalled, by the telephone companies. The telephone companies have flipped-flopped back and forth when it suited them. The claim that has been made for a number of years is that local telephone rates are priced below costs and benefited from a subsidy from long-distance revenues. These assumptions and practices are based on an arbitrary assessment by the telephone companies that has never been subject to an independent assessment by the CRTC or others. Consequently, critical communications researchers and progressive consumer and public-interest groups have disputed the local-subsidy myth (Babe, 1990: 121–6; Mosco, 1990b; Rideout, 1993: 37; Roman, 1990: 96–110; Tinker, Leham, and Neimark, 1988: 188–216).

In a news release responding to the CNCP application, Bell stated that it was in fact prepared to compete in the long-distance telephone market, providing the telephone rate structure was changed (Bell, 1983). Bell was adamant that if competition was allowed in long-distance services, then local and long-distance rates would have to move closer to costs, otherwise known as rate rebalancing. Under rate rebalancing Bell advanced its idea of the total elimination of the subsidy to local rates. This was the second time Bell had brought forward its rate-rebalancing scheme. The first, it will be recalled, was in its brief to the Macdonald Commission. In essence, Bell supported long-distance competition but threatened the CRTC, CNCP, and its telephone subscribers that if a competitive long-distance market was implemented, the rates for local and rural and remote services would rise dramatically. Long-distance rates, by contrast, would fall for heavy long-distance users. Further, Bell proposed the implementation of local measured service (LMS). If LMS was approved, Bell could charge local subscribers based on their telephone usage. Every time a subscriber made a local telephone call, they would be charged for the call based on the number of minutes of usage. Similar forms of LMS were already being used in Europe, including the United Kingdom, and in some parts of the United States. Some have argued that the reason telephone penetration rates have increased throughout the twentieth century in Canada and the United States is because all the telephone companies offered basic service at a flat rate rather than a measured one (Hills, 1986).

CONSUMER AND PUBLIC-INTEREST POLICY RESISTANCE

Consumer resistance in Canada began with the formation of a national and provincial women's groups in 1947. Initially known as the Canadian Association of Consumers (CAC), the organization was recruited by the Consumers Branch of the federal Wartime Prices and Trade Board to assist in monitoring prices and rationing. Primarily the organization is made up of volunteer provincial and territorial CACs, and includes the national organization based in Ottawa. The national organization studies consumer problems, makes recommendations, and brings the views of consumers to the attention of government, trade, and industry. By 1961 the CAC had broadened its mandate from strictly women's issues to public-interest issues and changed its name to the Consumers' Association of Canada (CAC, 1986: 1–3). Beginning in the mid-1980s the CAC took on a more pro-active advocacy role and began participating in public telecommunications hearings at the CRTC.

At the first long-distance competition hearing (CRTC 85b–19) the CAC was opposed to rate rebalancing, long-distance competition, and local measured service because it was thought that competition would not benefit consumers. At that time the CAC was still concerned about the adverse impact such a policy change would have on low-income households, the elderly, and non-profit organizations, which make extensive use of local telephone services (CAC, 1986: iii). The CAC opposed CNCP's application mainly because it was not convinced that lower long-distance rates would accrue to all residential subscribers. According to the CAC, a more likely scenario was that competition in long-distance phone services would result in higher local rates (ibid., 28).

It would appear that the CAC was, at the time, concerned about all telecommunications consumers, including the economically disadvantaged. But internal reorganization within the CAC and the creation of the Regulated Industry Program (RIP) moved the organization away from the anti-competitive position it shared with other consumer and public-interest groups to a more ambivalent position regarding competition and telecommunications. According to the former director of the RIP, David McKendry, this program was developed in 1980 to produce sophisticated research that would address the interests of consumers in a competitive environment. The CAC was opposed to the 1985 CNCP application because the corporation did not provide any evidence that consumers would be better off financially if long-distance competition were introduced. In addition, evidence was presented at the hearing that local rates would go up (interview with D. McKendry, March 1997). Even though the RIP was eliminated in 1989, the CAC, under the executive direction of one of the former RIP lawyers, Rosalie Daily-Todd, moved away from supposedly representing all consumers to represent only middle-income and small-business phone subscribers at the second long-distance competition hearing.

The National Anti-Poverty Organization (NAPO) was also strongly opposed to CNCP's application. NAPO was founded in 1971 at Canada's first nation-wide poor people's conference. It is directed by activists working with low-income communities. Its twenty-two-person voluntary board of directors is drawn from every province and territory of the country, and all board members either are living or have lived in poverty at some point in their lives. NAPO has more than seven hundred individual and organizational members[2] (NAPO, 1997).

The organization has been one of the most actively involved social organizations in CRTC telecommunications hearings since 1977. NAPO's main concern then was that policy liberalization would result in basic local telephone-rate increases. NAPO noted that CNCP's application lacked evidence to support its claim that local rates would not increase. In NAPO's view the existing telephone system was a natural monopoly, with known rates and cost advantages that were based on economies of scale and scope. This would mean that a small competitor like CNCP would not be able to compete without CRTC-imposed concessions or business subsidies (CRTC 85b–19, 1985: 29). NAPO, the Consumers' Association of Canada (CAC), the Federated Anti-Poverty Groups of British Columbia (FAPG et al.), Hurontario, and both telecommunications unions considered CNCP's application to be financially uneconomic. The consumer and public-interest groups and the governments of Quebec and the Atlantic provinces, as well as the CFCW and the Telecommunications Workers Union, expressed grave concern about the impact that long distance competition would have on the principle of universality for basic service (ibid., 29–34).

TELECOM TRADE-UNION POLICY RESISTANCE

In addition to participating in the Interexchange Competition hearing (CRTC 85b–19) to voice their concerns over universality and potential job loss, two unions, the CWC and the TWU, were very active in informing the Canadian public about telecommunications competition and deregulation issues. The CWC launched a national campaign about the negative aspects of long-distance competition in both official languages. By advertising on the major public and private radio and television networks, placing ads and articles in many of the major daily newspapers, and providing information leaflets, the CWC presented the other side of the telephone-completion story, including the likelihood that local telephone rates would increase, that telephone quality of service would be eroded, and that many telecommunications jobs would be lost (CWC, 1984: 1). In addition, the CWC was also instrumental in a grassroots campaign that resulted in the Department of Communications receiving more than 10,000 pieces of mail decrying CNCP's application (CWC, 1984).

The Telecommunication Workers Union submission to the CRTC hearing cited numerous examples of telecommunications competition problems in the United States, including rate increases in California, Idaho, Montana, and Texas that ranged from 50 to 201 per cent, the erosion of quality of service, and dramatic increases in the costs of installation and repair services (TWU, 1984: Appendix 1). The TWU continued to support a monopoly-controlled end-to-end regulated telephone system. In a cautionary note, the TWU explained, "the Canadian telephone companies are not really serious in fighting competition, competition will allow them to get out from under the obligation to provide universal service at an affordable price" (TWU, 1984: 1). Previous evidence and the TWU's comments reveal that Bell and BC Telephone were only committed to the principles of universality and affordability providing their monopoly status was maintained.

CNCP's first effort to enter and compete in the long-distance telephone market was denied by the CRTC because it "would not be in the public interest" (CRTC 85b–19, 1985: 43). The commission did agree with CNCP, the user lobby organizations, and the Director in their support for the principle of competition. However, the CRTC did not believe that CNCP would be able to offer significant benefits throughout the territories served by Bell and BC Telephone (ibid., 45). Moreover, the CRTC was not convinced that CNCP would be able to meet the legislative requirements to provide universal service. Closer analysis reveals that the commission's rationale for denying CNCP entry into the long-distance market was primarily based on a financial assessment of the company's profitability and its projected shareholder return. The CRTC pointed out that CNCP would operate at a loss for the first four years. Over the next six years it projected that it would provide its shareholders with a rate of return of 21 per cent[3] (CRTC 85b–19, 1985: 45). As the Peat, Marwick and Partners study conducted for the Department of Communication noted, these projections were definitely not reasonable and very unrealistic (ibid., 34).

Both telecommunications unions (CWC, TWU), the public-interest advocates for the national (NAPO) and British Columbia (FAGP et al.) anti-poverty organizations, as well as the municipal government of Hurontario and the provincial governments of the Atlantic region, Quebec, Manitoba, and Saskatchewan were successful in defeating the first bid at long-distance telephone competition. This success, however, was short-lived. As soon as the CRTC denied entry liberalization in public long-distance telephone service, a number of pro-competition agents initiated new strategies, held conferences, conducted research, and delivered policy position papers and public statements promoting the necessity of liberalizing long-distance services (Stanbury, 1986). No sooner had the CRTC denied CNCP's application than one of the federal state agents, DOC, became more active to further liberalize the telecommunications environment.

FEDERAL GOVERNMENT INTERVENTION

In August 1985, less than a week after the CRTC issued its decision denying CNCP entry into the public long-distance telephone market, the minister of Communication, then Marcel Masse, announced that the government would conduct a federal telecommunications policy review. One of the major complaints from many of the provincial governments that participated in the Interexchange Competition hearing, whether they did or did not support CNCP's application, was that the provinces were not being consulted concerning this major policy change.

According to a former Department of Communication employee, the objective of the policy review was to forge a policy consensus (confidential interview, 1997). But a leaked DOC memo to the federal Cabinet shows that consensus-building was not very high on the department's agenda. Rather, the federal government was proceeding in an extremely divisive way, attempting to fractionalize the small users, the provinces, and other social organizations.

The DOC memo recommended that a consultation process be undertaken to inform the public about important telecommunications issues. Not only did the DOC support liberalization policies; it stated that a move to increased competition would require a new telecommunications policy. The memo conceded that competition should be introduced gradually, "as conditions permitted" (DOC, 1985: 1). What was most telling, however, was that the DOC recognized that "a new telecommunications policy could be considered as favourable to big business. Therefore, the government must develop a strategy to dissipate the formation of a common front between the public, the provinces, and the PME [*petites moins enterprises*]"[4] (DOC, 1985: 2).

To help persuade these policy agents, the federal government was prepared to spend one million dollars on a public-awareness campaign that included public-opinion polls (DOC, 1985: 11). First of all, the memo stated, it was inevitable that rates would have to be "rebalanced," a Canadian "snappy euphemism," according to Mosco (1989: 213), which would result in lowering long-distance charges and raising local rates. The leaked memo noted that such a move would not go over very well with the majority of Canadian telephone subscribers. In fact, most Canadians were satisfied with the telephone system, so it was up to the federal government to prove that these changes were necessary (ibid., 9). Such a move worried the federal government: "Given that consumers, small business and provinces will join to fight rate restructuring, the Government must prevent at all costs the formation of a common front by launching a major consultation program. Such a common front would create the biggest threat to the Government's ability to manage the changes" (DOC 1985: 5). DOC was concerned that if a common front succeeded, it would implicate the state in changing telecommunications policy to benefit big business regardless of the outcome for everyone else.

The DOC memo also recommended that a national debate be held to review telecommunications policy. Two documents contributed to this national debate, the first prepared by Communications Canada (DOC), the second by the CRTC. *Communications for the Twenty-first Century* (DOC, 1987) claimed to contribute to the national debate on communications; since this national debate was not open to the public, one must surmise that the debate included only federal and provincial bureaucrats, perhaps the telephone companies, the new telecommunications competitors, and perhaps the telecommunications-user organizations. The policy section of the DOC report focused on advancing competitive telecommunications policies and continentalization. It referred to a 1987 federal government policy statement that placed special emphasis on telecommunications and clarified the rules for the new telecom arena. The DOC report distinguished between two types of carriers, those that own the basic infrastructure and those that use the infrastructure to provide value-added networks (VANs – enhanced) service. The first type (Class I carriers) were to be governed by the recognition that a telecommunications infrastructure is an essential component of national sovereignty and therefore must continue to be 80 per cent Canadian owned.[5] Other requirements for Class I carriers included full available interconnection with all other telecommunications competitors (DOC, 1987: 56–7). The second type of carrier, Class II, would include both VANs and authorized resellers. Class II carriers would rent transmission capacity from Class I carriers and then augment these facilities. Policies governing these services were changed to allow open, unrestricted entry for all competitors and reliance on market regulation. In Communications for the Twenty-first Century DOC also advanced telecommunications continentalization by lifting the foreign-ownership restrictions for all Class II carriers (ibid., 57). Lifting these rules for telecommunications business services was particularly sweet for the user organizations (the multinationals), which conduct their business in the North American continent and internationally. This pro-competition policy move also proved to be very rewarding for the many Canadian and U.S. VAN operators and resellers.[6] A continental standard was adopted for basic and enhanced services (including value added) that would have an important impact on the Internet, multimedia, and electronic commerce.

As the report indicated, the federal government was busy announcing telecommunications policy changes that would further entry liberalization, competition, and continentalization. The CRTC, by contrast, had the task of defusing explosive domestic issues such as deregulation and rate rebalancing. The federal government dealt with provincial telecommunications jurisdiction problems.

The DOC document also stated that it was important to deflect discussions with the provinces regarding economic objectives in order to avoid the conflict that was erupting over telecommunications jurisdiction. The federal government was already aware that some of the provinces, including Ontario and British Columbia, supported competition (CRTC 79–11 and 85–19). The conservative government of Alberta was prepared to consider competition. But the governments of Saskatchewan and Manitoba were not, because competition would require rate rebalancing. Quebec did not take a position, and the governments of the Atlantic provinces were convinced that telecommunications policy needed to be modified but were worried about the political consequences of moving to competition and rebalanced rates (DOC, 1985: 7).

The Federal-Provincial-Territorial Task Force on Telecommunications (Sherman Report), which was carried out by the CRTC, focused entirely on economic issues. It began by asking policy questions. The major question it addressed was whether competition would benefit long-distance telephone services, or whether monopoly long-distance service should continue (Sherman Report, 1988: xxix). Before analysing the report, it is important to consider just how this question was dealt with, since it established the parameters under which the task force and public debate took place. The only participants, however, were bureaucrats, and the federal, provincial, and territorial ministers responsible for communications. The Sherman Report even stated that the general public, which may have been interested in the long-distance issue, were not invited to participate. This resulted in two major problems with the report. First, information and sources could not be challenged or verified. Second, this exclusion of the general public from matters of public interest would continue in the International Trade Advisory Committee, the Canada–u.s. Free Trade Agreement, and the North American Free Trade Agreement.

The task force explored three telecommunications competition models: a fully regulated market model; an intermediate model involving the regulation of only the dominant carriers; and an unregulated, competitive market model (Sherman Report, 1988: xxxiii). The fully regulated market model would be one in which entry liberalization would be permitted but all market participants would be regulated. Benefits would include very little if any disruption in rate movements and the ability of regulators to control all aspects of entry, exit, and service provision (ibid., xxxiv). In the intermediate model, dominant carriers would be defined as the service providers who are able to exercise market power. In contrast, non-dominant suppliers would be unlikely to exercise market power. Under this model, if long-distance competition were permitted, the telephone companies would continue to be regulated, while new entrants would be free from regulation. Regulators would also have the means to impose on dominant carriers the obligation to serve, entry and exit criteria, and quality-of-service requirements. In the unregulated, competitive market model all carriers would be completely unregulated – that is, the incumbent telephone companies and the new entrants would be free to compete without regulatory controls except for local telephone service, which would continue to be regulated. In such an environment there would be great pressure to move towards competitive market pricing or rate rebalancing (ibid., xxxiv–xxxv).

The Sherman Report's own summary claimed that, after considering the technical, operational, and economic factors introduced in its various studies, the task force had found no conclusive evidence to settle the issue of whether public long-distance telephone service should be provided on a monopoly or a competitive basis (ibid., 118). A closer examination of the report reveals that the solution that was being advanced for the long-distance telephone problem was a move to one of the competition models. In each model, competition is the constant. The only change that each model offered was varying degrees and ranges of regulation. Despite its attempt at an appearance of neutrality, the task force was really a subtle endeavour at consensus-building, to get the provincial and territorial governments to buy into one of the competition models. But when the ambiguous statement of neutrality from the Sherman Report was combined with the CRTC's 1985 decision denying CNCP entry into the long-distance telephone market, the large telecommunications-user organizations were not pleased. Both the task force and the Interexchange Competition decision intensified pro-competition political activism throughout the next six years.

ACTION BY PUBLIC AND PRIVATE RESEARCH INSTITUTES

The first research institute to engage in neo-liberal telecommunications policy activism was the Institute for Research on Public Policy (IRPP). IRPP was a big endorser of liberalization and deregulation. After the CRTC denied CNCP's entrance to compete in the long-distance telephone market, the institute sought the views of a number of pro-competition organic intellectuals, such as S. Globerman, H. Janisch, W.T. Stanbury, B. Woodrow, and K. Woodside, to conduct a post-mortem on the 1985 decision. The resulting collection of essays, *Telecommunications Policy and Regulation: The Impact of Competition and Technological Change*, was a major critique of the interexchange decision.

The central theme of the various contributors was the inevitability of increased telecommunications competition. It was noted, however, that increased competition would create winners and losers. The winners, the large corporate users, would be able to benefit from substantial savings on long-distance calling. Residential and small-business users, however, would be the big losers, paying more for local services. Even though Janisch painted this bleak picture for local residential telephone subscribers, he was adamant that failure to embrace telecommunications competition would result in the large users bypassing[7] the Canadian network (1986: 309–10). He explained that many of the large users are subsidiaries of American firms and already familiar with the latest telecommunications services and networks that their parent companies use.

Other contributors such as Woodrow and Woodside agreed that telecommunications competition was inevitable. They went on to explain that the process involved bringing about a policy shift based on a complex relationship of technological change, involving pressure from a number of telecommunications players, and political factors (1986: 101–2). Woodrow and Woodside provided a comprehensive list of the players and the stakes, and identifed major political factors, but provided little in the way of concrete analysis. Rather, their research concentrated on economic factors and technological advance, including the competitive drivers that would push the price of telephone services closer to costs. The introduction of competition would require corresponding regulatory policy changes, from monopoly regulation to managed regulation (238). But by concentrating on technology as the sole force of change, Woodrow and Woodside presented an oversimplification of the telecommunications issues. Their analysis amounted to a technological-deterministic assessment of telecommunications policy and regulatory changes, one that did not take into account the complex socio-political relations involved in the interexchange hearing.

Other IRPP contributors such as Stanbury chastised the CRTC for making the wrong decision in 1985. Stanbury's assessment is closer to that of research conducted at the Fraser Institute. His main criticism was that Telecom Decision 85–19 was too political (1986: 508) – that is, that the commission had continued to protect the established stakeholders, the telephone companies, as well as the poor and disadvantaged. According to Stanbury, the criteria that should have been used to assess the case were economic rationality and efficiency (509). Stanbury concluded that local services had been subsidized by policies that were far too generous. His argument breaks down, however, because it was not the subsidy per se that Stanbury and the IRPP contributors objected to, but who should get it. What was being advocated was the subsidization of the heavy long-distance users by local subscribers. Others such as Richard Schultz, argued that the federal government should do away with all aspects of universal cross-subsidies, what Schultz refers to as "a hidden welfare system" (1995: 279). Schultz's argument was that the cross-subsidy from long-distance to basic local service helped the incumbent telephone companies rather than providing low-cost residential telephone rates.

The arguments made by Schultz, Stanbury, and the other IRPP contributors can be criticized on a number of levels. Although I agree that keeping the cost of residential services low via a flat rate helped the incumbent telephone companies, it also benefited Canadian consumers by providing them with universal access to basic telephone service, as the 98 per cent country-wide penetration rate indicates. Second, a lot of ambiguity remains about the cross-subsidies that are still in the telecom system, ambiguity that can be traced back to Bell Canada's early accounting practices and which were adopted by successive government regulatory bodies. The move away from rate rebalancing to cost-based pricing has removed some of the cross-subsidies, but as Commissioner McKendry explained, the whole telecommunications system still contains cross-subsidies (interview, September 2000). The neo-liberal telecom policy changes have proved difficult for the CRTC. Its approach has been to move gradually and manage the competitive landscape and narrow the cross-subsidies as much as possible.

Telecommunications in Canada: An Analysis of Outlook and Trends (1988) by S. Globerman, a study conducted for the Fraser Institute, also supported the neo-liberal policy platform. Globerman identifies two domestic policy issues, distributional equality and deregulation. The first issue is domestic in nature, and it too raises the subsidization issue. Globerman argues that general local subsidies should be replaced with a more efficient, targeted subsidy progam. Such a progam could target low-income subscribers through a welfare allowance, providing a minimal basic level of service. All other local telephone subscribers would be subject to a usage-sensitive pricing scheme, similar to the local measured service proposed by Bell Canada. Globerman had explained in his chapter in the IRPP study that subsidies are an "allocative inefficiency" that amounts to two billion dollars a year, one that supposedly goes from long-distance service to local (1986: 123). The second issue, deregulation, has an urgency attached to it because of its international implications. If the policies affecting long-distance competition are not changed, large business customers will be forced to bypass the common-carrier facilities to use private networks, encouraging instead cross-border bypass to the United States or international reseller bypass (122).

By concentrating on the continental aspect of telecommunications relations, the IRPP and Fraser Institute continuously reminded the federal government and others of the implications of a telecommunications bypass. Bypass, then, could almost be construed as a threat to the federal government from the large corporate users as a way to hurry up regulatory reform.

TELECOMMUNICATIONS-USER ORGANIZATIONS

The oldest telecommunications-user organization, the Canadian Business Telecommunications Alliance (CBTA), was formed in 1962 to act on behalf of business users to bring their views to the attention of Canadian policy-makers (Annual Report, 1995: 20). More than four hundred enterprises from all Canadian business sectors and many public-sector institutions make up the CBTA membership (see Appendix 13). The organization's strategies and objectives are decided by its telecommunications public-policy committee. Table 5.1 reveals that this committee is predominantly made up of Canadian and U.S. multinationals in banking and the computer industries. Many of the CBTA policy-committee members are also members of other business lobby organizations, such as the BCNI, among others (see Appendix 1).

CBTA strategies have included support for more consumer choice, innovation, quality responsiveness, and cost effectiveness in all telecommunications markets (Annual Report, 1995: 21). Although the CBTA supported competition and choice in the first long-distance competition hearing, it supported both CNCP's application and Telecom Canada's rate-rebalancing proposal (CBTA Bulletin, 1988: 2). It is clear that the CBTA threw its support behind both providers in the hope of getting long-distance rates lowered for its members. According to the CBTA, its members' combined telecommunications expenditures were more than $4 billion a year. In one of its position papers, *Competition the Future for Canadian Telecommunications*, the CBTA stated that its primary goal was to enhance competitiveness (1991b). It promoted competitiveness through the revenues it collected from members to conduct research and to participate in public telecommunications hearings. These revenues ranged from $1.3 million per year in 1989 to $2.3 million for 1992, ample resources to promote the pro-competition view proclaimed in its position paper: "Business users must have available to them the widest possible range of telecommunications services at the lowest possible prices. This means unrestricted choice of a broad range of innovation services and facilities at prices equal to better than the best in the world" (1991b: 1). And, according to the CBTA and a newly created telecommunications-user organization, Canadians for Competitive Telecommunications (CCT), the best long-distance telephone prices in the world were just south of the Canadian border, in the United States.

Large telecommunications-user activism involved the continuous recreation of some members into additional user organizations. These user organizations, however, were, to paraphrase Shakespeare, just roses by other names. The Canadians for Competitive Telecommunications (CCT), formed in 1986, was one of these recreations. Appendix 11 lists the members of the organization. Members include the Canadian Business Equipment Manufacturers' Association (changed to the Information Technology Association of Canada), the Canadian Business Telecommunication Alliance, and the Canadian Federation of Independent Business. All were actively involved in the first failed long-distance competition hearing.

In its 1986 report *The Crisis for Canadian Business: Telecommunications and the Public Interest* the CCT painted a very bleak picture for Canadian business unless long-distance competition policy is introduced as quickly as possible. The report pointed out that the third-highest operating costs for many Canadian businesses were telecommunications costs (7). Different organizations including the CCT provided evidence that showed that Canadian long-distance telephone rates (Telecom Canada members) were 50 to 100 per cent higher than those of AT&T, MCI, or Sprint (8–10). The report went on to note that

Table 5.1
CBTA Telecommunications Public Policy
Committee members as of 1994–95

Bank of Montreal
BC Systems Corporation (Government of BC)
Canadian Bankers Association
EDS Canada (data processing)
Government of Northwest Territories
Hewlett-Packard
Hewlett-Packard (Canada)
IBM Canada Ltd.
Management Board Secretariat CTS (Government of Ontario)
· Nesbitt Burns
Ontario Hydro
Royal Bank of Canada
Regional Municipality of Peel
Southam Information
State Farm Insurance
Telecom Division (data processing)
Thomson Newspapers
United Parcel Services

Source: Adapted from CBTA Annual Report, 1994–95

many CCT members were multinationals who participated in a continental and international marketplace. According to the CCT, in order
for these businesses to compete in this market, they needed access to
long-distance telephone rates that were compable to those in the
United States (11).

In 1987 the Canadian Business Equipment Manufacturers' Association (CBEMA) renamed itself the Information Technology Association
of Canada (ITAC).[8] Formed in the 1950s, CBEMA was initially a trade
association for business- and office-equipment suppliers. Towards the
end of the 1970s the organization intensified its political activism,
supporting pro-competition policies and neo-regulation. By 1991
another user organization, the Canadian Association of Data and Professional Software Service Organization (CADPSO), merged with ITAC.
CADPSO was founded in the 1970s to represent the interests of the
large-scale computer-service industry (ITAC, Backgrounder). Recalling
that ITAC is also cross-affiliated with the BCNI executive, Appendix 12
lists some of ITAC's 450 members. ITAC is also affiliated with six
partner organizations in British Columbia, Ontario, Quebec, New
Brunswick, Nova Scotia, and Newfoundland. As Table 5.2 shows,
ITAC's board of directors is predominantly made up of executives from
Canadian and U.S. multinational corporations involved in telecommunications, computer hardware and software, data, and cable industries.

Table 5.2
ITAC Board of Directors 1993–94

3M Canada Inc.	Lotus Development Canada Ltd.
AT&T Canada and Global (2)	Metcan Information Technologies Inc.
Alias Research	Microsoft Canada Ltd.
Auto-Trol Technology Canada Ltd.	Nortel North America
Apple Canada Inc.	(Northern Telecom – NorTel)
Anderson Consulting	Oracle Corporation Canada
Ambrex Technologies Inc.	Pitney Bowes of Canada Ltd.
Atlantic Computer Institute	PMP Associates Inc.
Brant Interprovincial System Inc.	Rogers Communications Inc.
Bull Information Systems Ltd.	SHL Systemhouse
CGI (Group) Inc.	Siemens Electric Ltd.
Compacq Canada Inc.	Silicon Graphics Canada Inc.
Comshare Ltd.	Sierra Systems Inc. (2)
Data General (Canada) Inc.	Software Kinetics Ltd.
Digital Equipment of Canada Ltd.	Southam Newspapers
DMR Group Inc.	Softworld '94
EDS Canada	Stentor Canadian Network Management
Elan Data Makers	Stentor Telecom Policy Inc.
Evans Technologies Inc.	Stentor Resource Centre Inc.
Hewlett-Packard (Canada) Ltd.	Sun Microsystems of Canada Inc.
IBM Canada Ltd.	Tandem Computer Canada Ltd.
ISM-BC Information Systems	Unisys Canada Ltd.
IST Computer Service Co.	Unitel Communications Inc.
ITAC Ontario	Wang Canada Ltd.
LGS (Group) Inc. (2)	Xerox Canada Inc.
Linktek Corporation	

Source: Adapted from ITAC, *Annual Review,* 1993.

Its political advocacy includes government information-technology policy positions, free trade, and international policy development. What is different about this organization is that ITAC policy positions are not always arrived at on a consensus basis (interview with Bob Crow, ITAC, 1996).

In the ITAC report *An ITAC Statement on Competition Policy for the Telecommunications Industry (1990)*, the organization supported a telecommunications market where no single buyer or seller would be powerful enough to dictate what goods and services would be offered, their prices, and/or under what conditions and terms they would be offered (2). This position did not sit very well with three of the association's board members, Stentor, BC Telephone, and Bell Canada. Attached to the report are disclaimers from Bell and BC Telephone. Bell did not agree with any of the views expressed in the report. Taking a more diplomatic approach, BC Telephone stated that it held views similar to those expressed in the report but believed that the proposed policy goals could be achieved using the existing telecom infrastructure (5). A majority of the board members, however, supported entry liberalization and deregulation for other telecommunications competitors. Unitel and Rogers, Bell and BC Tel's competitors, supported open competition provided that the telephone companies would continue to be regulated (3).

Yet another reincarnated business advocate, the Communications Competition Coalition (CCC), was formed in 1989 specifically to lobby for a single issue, long-distance competition. Appendix 14 shows that most CCC members were already members of other telecommunications-user organizations such as the CBTA, ITAC, and the Canadian Bankers' Association. The CCC was even more elitist than the others. Membership was restricted to the CEOs of multinational corporations in finance, manufacturing, resources, transportation, and the mass media, much like the make-up of the BCNI, but with an annual membership fee of $50,000 (Surtees, 1994: 192).

Strategies for the CCC were formed by a six-member steering committee that included a vice-president from the Royal Bank, a former vice-chairman of the CRTC, the lawyer for Rogers and Call-Net, H. Janisch, a law professor at the University of Toronto, the president of the Gemini Group, the deputy chairman of Rogers Communications, and Monty Richardson, the executive director for the CCC. Their strategy included a major lobbying campaign and raising resources ($2.5 million) to participate in the next long-distance competition hearing (Surtees, 1994: 232). The CCC's mandate was to make sure that "true competition" prevailed in Canadian telecommunications policy. Even though the CCC supported Unitel's (CNCP Telecommunications and Rogers Communications) application to enter the long-distance market, the coalition did not support the duopoly model advocated by Unitel. What the CCC supported was total open competition in long-distance telephone services (interview with Monty Richardson, May 1996).

One of the largest telecommunications users, the Royal Bank, engaged in direct political activity to bring about policy liberalization and regulatory reform. In a meeting at the Canadian Club in Toronto in the fall of 1989 the then Royal Bank CEO, Allan Taylor, delivered two messages to the federal government: "First we need one national telecom jurisdiction; and second we need competition within it ... We need a new telecom policy in which choice is the corner stone" (A. Taylor, Royal Bank 1989: 185, 189).

Taylor's message was extended in a study commissioned by the Royal Bank, *Exploiting the Information Revolution: Telecommunications Issues Options for Canada (1989)*, where in the bank stated that the monopoly that had dominated the telecommunications industry in Canada had outlived its usefulness. The study was a call to arms for business users. It advocated that the telecommunications environment should be controlled by business users, not the telephone monopolies. Consequently, these users needed to play a greater role in telecommunications policy-making, advancing policies that liberalized entry for new telecommunications providers at both the domestic and international levels (1). At the domestic level the bank supported policies that would permit open competition in the long-distance telephone market, not a duopoly (35). The report also supported only a light form of regulatory oversight in the transition stage to a competitive environment. Once the transition was completed, the report recommended, the Competition Bureau and competition policy could oversee the telecommunications environment while the Director would work towards further reducing the role of traditional regulation (35).

A TELECOMMUNICATIONS CONFERENCE FOR ALL CANADIANS

Yet another rallying cry for policy liberalization came from the telecom-user lobby organizations and the new telecommunications challengers. It occurred at a conference on long-distance competition held in Toronto in 1989. Although some Telecom Canada members were invited to participate, the conference provided an overwhelming endorsement for an end to the telecommunications monopoly and the introduction of competition in the provision of long-distance services (Cruickshank, 1989). Only two participants were invited who had alternative views, the British Columbia Public Interest Advocacy Centre and the government of Saskatchewan.

It was another conference that encouraged public debate on telecommunications policy. "A Telecommunication Policy for All Canadians" brought together experts from universities, government, business, labour, consumer groups, and the media to discuss the problem and prospects for Canadian telecommunications. A first of its kind, the conference was co-sponsored by Carleton University and the Communication and Electrical Workers of Canada (the successor to the CWC), and organized by the organic intellectual Vincent Mosco. The major issues of the conference included universality, affordability, and lessons from the American experience. Social importance, rather than economic issues, framed discussions on universal access to basic telephone service for urban/rural subscribers and residential/business subscribers (G. Lang, Minister of Communication, Saskatchewan, 1990; Carleton Media and Communications Research Centre, 1990). Additionally, the conference highlighted some of the important lessons from the United

States regarding long-distance competition and divesture, and their effects on local rates. In just five years, local rates rose by 43 per cent, whereas long-distance rates dropped by 35 per cent. This dramatic rise in local rates disproportionately affected the poor and low-income peoples, as well as black and Hispanic Americans (Mosco, 1990: 14; Consumer Federation of America, 1990; Carleton Media and Communications Research Centre, 1990).

The conference was successful in opening up debate about the major changes taking place in Canadian telecommunications policy. Many of the participants at the conference, such as the unions, academics, and consumer groups, can express their alternative views and positions only occasionally in public forums such as this one because of their limited resources and personnel. Consequently, these individuals and groups were unable to mount and maintain ongoing pressure to alter the new policy course. The telecommunications-user organizations, by contrast, such as the CBTA, ITAC, and the CCC, had access to an inordinate amount of financial[9] and human resources. Despite this array of user organizations with multinational pedigrees, the contest for hegemonic continental consent only intensified the alternative hegemonic forces.

THE CENTRALIZATION OF
TELECOMMUNICATIONS REGULATION

The Supreme Court ruling *Alberta Government Telephone and the* CRTC *and* CNCP *Telecommunications el al.* (1989) was another aspect of federal state agency that was necessary to bring all telecommunications under federal regulation before more extensive policy liberalization could be carried out in Canada. Again, as they evolved after Confederation, telecommunications jurisdiction and regulation were very fragmented,[10] spread among federal, provincial, and municipal levels of government. By the mid-1980s this divided jurisdiction had become troublesome for new challengers such as CNCP and the telecommunications users. When the CRTC granted CNCP permission to interconnect its private-line voice and public data services to the public switched telephone network, that permission applied only to areas under CRTC jurisdiction – that is, British Columbia, Ontario, Quebec, and two northern territories. This was also the case for terminal equipment and business services.

In 1982 CNCP Telecommunications applied to the CRTC to interconnect its private and data network services to those of Alberta Government Telephone (AGT). The CRTC sent the request to the Federal Court of Appeal because of its jurisdictional and constitutional implications. By 1984 the Federal Court ruled that telephone companies who are members of Telecom Canada constitutionally fall under federal jurisdiction because they offer interprovincial phone service. This meant they would also be subject to regulation by the CRTC under the Railway Act. AGT appealed the decision to the Supreme Court of Canada. Five years later the Supreme Court upheld the Federal Court ruling, agreeing that provincially regulated telephone companies fall under federal jurisdiction (Supreme Court of Canada, 1989).

In this landmark ruling the Supreme Court stated that each Telecom Canada member operated a single integrated interprovincial undertaking, making them subject to federal regulation by the CRTC. As the ruling indicates, each of the provincial telecommunications companies in the Atlantic and prairie provinces was now subject to CRTC regulation. Alberta Government Telephone was privatized in 1990 and was immediately subject to federal regulation. In an agreement reached with the federal government, the regulation of Manitoba Government Telephone was transferred to the CRTC in 1993.[11] Although the government of Saskatchewan was initally strongly opposed to federal regulation the CRTC agreed to a transitional approach to regulation as Sasktel fell under CRTC jurisdiction on 30 June 2000 (CRTC 2000a–150). Sasktel and the CRTC agreed to an 18-month transition period from the time the decision was issued, to allow the company time to align itself with the current regulatory framework of the rest of the telephone companies.

This structural reorganization of telecommunications jurisdiction and regulation centralized power and control within the federal government, away from the decentralized telecommunications environment that had previously been shared among federal, provincial, and municipal governments. For important areas such as telecommunications, neo-liberal policy changes also require centralization and strong federal state action rather than decentralized regulatory approach. Consequently, the Supreme Court decision provided the control and uniformity necessary in the telecommunications sector to advance liberalization, neo-regulation, and the expansion of continental free trade. This centralized regulatory approach works in conjunction with other neo-liberal policy changes by effectively shifting control and power away from the municipal and provincial governments, who, in the past, had shown more concern for the social or political aspects of telecommunications than for only economic ones. Essentially this structural reorganization created a one-stop regulatory environment for the large telecom users and the new telecom challengers. From here on, the large telecommunications users and all the telecommunications providers would, as far as policy and regulatory matters were concerned, have to deal with only one level of government rather than two, or in some cases three.

6 Consumer and Public-Interest Resistance

We speak for the majority of Canadians who will get only higher telephone bills from this decision [Telecom Decision 94–19]. We have had telephone service in Canada that is universal and accessible to almost everyone. The CRTC decision threatens the universality of this essential service without benefiting the vast majority of Canadians.
                                               Michael Janigan, PIAC/PATS, 1994

Most importantly 80 per cent of those Canadian homes with incomes of less than fifteen thousand dollars will experience higher telephone bills. Many people simply cannot afford such increases.
                                               Marie Vallée, FNACQ/PATS/CSTA, 1994

LONG-DISTANCE TELEPHONE-SERVICE COMPETITION

With its 1992 decision *Competition in the Provision of Public Long Distance Voice Telephone Services and Related Resale and Sharing Issues*, the CRTC moved towards telecommunications-competition management. It is noteworthy that, after numerous submissions, evidence, and expert witnesses, the final decision is written in such a way that it appears as if there was little political opposition other than that of the established telephone companies (CRTC 92–12). From this point on the commission rarely made concessions to the progressive forces that challenged the now-dominant neo-liberal position. The following assessment shows that the second long-distance competition hearing was also strongly contested.

In this liberalization hearing two challengers, Unitel Communications (CNCP Telecommunications and Rogers Communications) and BC Rail Lightel (Call-Net and BC Rail Telecommunications), applied to the CRTC to interconnect their networks to the public switched telephone networks[1] so that they could compete in the long-distance market. Appendix 15 provides a complete list of hearing participants as well as the presiding CRTC commissioners. The appendix shows that many of the participants had appeared at the first long-distance hearing, with the exception of two additional user organizations, the Canadian Bankers Association and the newly formed Communications Competition Coalition.

Unitel came well prepared with an extensive business plan and a phased roll-out on a region-by-region basis that would take place over a six-year period. The company planned to offer various pricing options that would provide long-distance rates that ranged from 15 to 35 per cent below those of the established telephone companies. Promises were made for the provision of efficient services at lower costs, rapid introduction and diffusion of new services and facilities, and that entry would not threaten affordable universal service. Unitel proposed that it be allowed a 25 per cent discount on its contribution payments for the first three years of operation (1993–95); 15 per cent in 1996; and 10 per cent in 1997 (CRTC 92–12, 1992: 28). Unitel's rationale for its contribution plan was that it expected to obtain only limited market coverage rather than to roll out a second national telecommunications network (84). BCRL, by contrast, wanted entry to interconnect its voice and data telephone-service networks to provide alternative telecommunications services to large corporations in the province of British Columbia (8).

Early in 1992, before the long-distance competition hearing, Telecom Canada was reorganized and renamed Stentor. Stentor was an alliance of the major telephone companies, including Bell Canada, BC Telephone, AGT, SaskTel, MTS, MT&T, NB Tel, Newfoundland Tel, Island Tel, and Telesat. According to Stentor, it moved to a more formal, complex organization in order to exercise its influence domestically, continentally, and internationally.[2] The organization's vision, strategies, and policy direction were set by the council of CEO's from its member companies. Decisions at the council were agreed to based on member consensus. According to G. van Koughnett, vice-president for Stentor legal and corporate affairs, the reorganized Stentor benefited from the synergy created by the council, greater efficiencies, and now had the ability to speak as one powerful organization, with one voice, at both domestic and international levels (Interview with Greg van Koughnett, 1997).

Stentor's three main components included the Stentor Resource Centre, Stentor Canadian Network Management, and Stentor Telecom Policy Inc. (see Figure 6.1). The Stentor Resource Centre (STC) acted as the centralizing agent for the member companies. STC was responsible for engineering, research and development, and marketing. Stentor Canadian Network Management (SCNM) managed and monitored all interprovincial networks within Canada and to the United States and Mexico. SCNM also administered the revenue-sharing plan for all the North American interconnections. Stentor Telecom Policy Inc. developed the strategies and policy positions as set out by the council (Industry Canada, 1996b: 43). Stentor Telecom Policy also represented

Figure 6.1
Stentor structure and decision-making as of 1996
Stentor Canadian Network Management

Souce: Adapted from information from Stentor Telecom Policy Inc.; G. van Koughnett, 1997; and Industry Canada, *The Canadian Telecommunications Service Industry,* 1996.

the members' interests in ongoing government relations, lobbying, and participation at the CRTC hearings, including the long-distance compe-tition hearing of 1992.

Stentor opposed Unitel and BCRL's entry to compete in the long-distance market because, in its view, a multiple-supplier environment would require higher production costs. According to Stentor these costs would inevitably be passed on to customers (CRTC 92–12, 1992: 19–20). The Stentor members also argued that Unitel's proposed levels of contribution would not be sustainable in a competitive environment and that rate rebalancing would be inevitable. Bell estimated that just to bring Canadian toll changes in line with those in the United States would require an additional line charge of $8.75 on top of the monthly local service charge (47).

Numerous submissions from pro-competition supporters included the Director's submission, including evidence from Robert Crandall[3] of the Brookings Institution, as well as membership surveys carried out by the Canadian Business Telecommunication Association (CBTA) and the Canadian Federation of Independent Businesses (CFIB). These submis-sions supported open-entry liberalization and long-distance competi-tion (the Director, 1991; CBTA, 1991; CFIB, 1991). The Director's report went further and recommended that liberalization be followed with a plan to deregulate Canadian telecommunications policy (Direc-tor, 1991: 1). The Canadian Bankers Association (CBA), in addition to participating through its membership in the Communications Compe-tition Coalition (CCC), also submitted their evidence in *American Long Distance Competition: The Power of Choice* (1991) (for a list of partial members see Appendix 17). The CBA took the position that Canadian telephone monopolies had outlived their usefulness and competition could only have positive effects for users. This report focused on the positive effects of competition in the United States, including lower telephone rates, volume discounts, and increases in telecommunications-shareholders' dividends (CBA, 1991: 6). Evidence submitted by the CCC showed that Canadian users wanted long-distance telephone-rate parity with the United States. The CCC even provided evidence that revealed that businesses operating in Canada paid anywhere from 1.8 to 4.5 times more for telecommunication services, including long-distance services, than did businesses in the United States (CCC, 1991: 19). However, a major research study conducted by the Organization for Economic Co-Operation and Development (OECD), the *Performance Indicators for Public Telecommunications Operators* (1990), revealed that, when a total basket of telecommunication goods[4] was compared on a country-by-country basis, Canadian rates for all business services (based on fixed and usage charges) were lower than prices in Australia, Britain, Germany, Italy, Japan, and the United States (51). Additionally, Canadian residential rates were 15 per cent lower than the OECD average, as was the price of international business calls (51, 60). Although this evidence was presented by the Communications, Energy and Electrical Workers of Canada, it appears that it was glossed over or entirely ignored by the CRTC. Certainly, the CCC's evidence was very selective, comparing long-distance rates between Canada and the United States only in the largest cities with the heaviest telephone traffic.

The Communication and Electrical Workers of Canada (CWC) and the Telecommunications Workers Union (TWU) both argued that allowing long-distance competition would not be in the best interest of the Canadian public. In its submission, the *Final Argument for the Communications and Electrical Workers of Canada before the Canadian Radio-television and Telecommunications Commission on Telecom Public Notice 1990–73* (1991), the CWC argued that long-distance competition would be very costly. Evidence that was presented by Mosco on behalf of the CWC revealed that U.S. federal and state governments had tried to pass the cost burden of competition on to residential and rural subscribers. The compromise that was reached was the replacement of universality with national and state telephone-welfare programs, namely Link Up America and Lifeline. These programs were essentially targeted-subsidy programs for the deserving poor. As the Mosco evidence shows, as of 1990, the programs were not very successful because they did not reach even a majority of the eligible subscribers. Link Up membership as of 1990 was a mere 3 per cent and Lifeline had only a 33 per cent subscribership (Mosco, 1991: 4).

Both the Federal Communications Commission (FCC) and the Mosco research demonstrate that, despite the federal and state subsidy programs, the United States had serious affordability and access problems based on income, geography, and race. These problems have continued to plague the country as it shifts to an information/knowledge economy. The Department of Commerce's National Technology and Information Administration (NTIA) has produced three reports in the second half of the 1990s, called *Falling Through the Net*, *I*, *II*, and *III*. These reports show which American households have access to telephones, to computers, and to the Internet, and which do not.

Findings on the telephone penetration rates reveal that from 1994 to 1998 there was only a slight increase, from 93.8 per cent to 94.1 per cent (NTIA, 1998: 2). The disparities that existed in the 1980s after the introduction of long distance competition have continued through the 1990s. For example, only 79 per cent of the households with the lowest income (less than $5,000) have telephones. This increases to 89 per cent for those with households that have incomes between $10,000 and $15,000. Tracking by race and ethnicity reveals significantly lower telephone penetration rate of 85 per cent for Blacks and Hispanics. One's income and where one lives also greatly affect whether people have basic telephone service or not (2–3). These same disparities of class, race/ethnicity, and region have an impact on access and affordability when it comes to electronic services such as personal computers and connection to the Internet (NTIA, 1999: 5–7). As the research from these reports indicates, there is still a serious telephone-service affordability problem in the United States. Moreover, there is evidence of a digital divide between those who have access to new PC technologies and electronic networks and those who do not. These differences are also widening because of affordability problems that are class/income based (9).

The TWU submission, *The Future of Canada's Telecommunication System: Canadians Have a Choice* (1990), explained that the new competitors were not interested in serving the whole Canadian population. Like the United Kingdom competitor Mercury, Unitel and BCRL were only interested in the lucrative telecommunications markets. TWU evidence presented by J. Hills showed that Mercury only rolled out its network in the city of London and some of the urban midlands, a far cry from the promise of a second national network. Consequently, the majority of residential, rural, and small-business subscribers received no benefits, neither lower telephone rates nor any improvement in access once competition was introduced (Hills, 1991: 5).

Both the National Anti-Poverty Organization (NAPO) and the BCOAPO et al. were very concerned about the effects so-called long-distance

competition would have on affordability and accessibility of local services. What made more sense to these organizations, considering the
geographic size of Canada and its relative small population, was the
continuance of a single telecommunications supplier, providing it was
adequately regulated. In NAPO's view this would prevent the cost of
duplicate facilities. Otherwise, the cost of multiple systems would lead
to increases in local telephone rates (CRTC 92–12, 1992: 42–3; interview with Lynn Toupin, director of NAPO, 1997).

The incongruous position of the Consumers' Association of Canada
(CAC) in this hearing was certainly at odds with the other consumer
and public-interest organizations. Although the organization claimed
to represent all Canadian residential telephone subscribers, in its submission, *Residential Telephony* (1991), the CAC took an ambivalent
position, claiming not to support the telephone monopoly or to oppose
competition. According to the CAC, it was looking for the best possible
deal it could get for consumers (1). This was a far cry from the position
the CAC had taken in the first long-distance competition hearing in
1985, or its position one year later in the report "Emerging Telecommunications Issues: The CAC Perspective" (1986).

In its 1991 submission the CAC provided evidence that focused on
the "total telephone bill" (3), the rationale being that people pay
telephone bills, not rates. According to the CAC, looking exclusively at
the basic monthly charge distorts policy-making because it disguises
the consumption patterns of residential telephone subscribers. Consumption of telephone service had been rising throughout the 1980s
in relation to declining relative basic prices. Consequently, for the CAC,
consumption patterns are identifiable once the components of local
service are broken down into basic local services, optional services,
and repair and installation (3). The CAC further stated that when the
consumer price index (CPI) was applied to the total telephone bill, the
telephone component of CPI rose by 19 per cent, whereas telephone
consumption throughout the 1980s had advanced to 27 per cent (5).
Using this logic, CAC then went on to show that the average telephone
bill represented only 1.3 per cent of household income for residential
subscribers. Countering its previous position in the first long-distance
competition hearing, the CAC stated that a total telephone bill needed
only to be stable, not to be affordable, and that residential telephone
subscribers, not just the large telecom users, should be able to take
advantage of lower long-distance telephone rates (2).

With its focus on consumption and the total bill, the CAC moved
towards selective representation of middle- and upper-class consumers,
consumers who make many long-distance calls and subscribe to an
array of optional services (44). This move by the CAC excluded its

poorer constituents on fixed or low incomes. The then CAC director of policy research, M. McCall, stated that the organization tended to support the other consumer groups on technical issues such as inside wiring but were not aligned with these groups regarding the costs to consumers for the total telephone bill and affordability issues such as local measured service and lifeline (interview with M.McCall, 1997). The CAC's advice to residential consumers, once long-distance competition was implemented, was to shop around and ask for savings from the equal-access providers such as Unitel and BCRL (Sprint Canada) or switch back to the Stentor companies if they were not happy with their telephone service (CAC, Tips, 1996).

The CAC's position on the total bill had a number of important repercussions. The CRTC praised the "total-bill" concept stating that the CAC had provided a useful measure for assessing affordability (CRTC 92–12: 49–50). Consequently, the basic portion of the telephone bill was unbundled to reflect basic service, optional services, and repair and installation charges. This in effect broadened the affordability measure for residential telephone subscribers and contributed to creating residential telephone haves and have-nots, based on individual consumption patterns. Redefining affordability came not only from the business-user organizations, the new competitors, and the CRTC but from a consumer organization that was supposed to represent all residential telephone subscribers, even though it had made its submission without any meaningful consultations with the other provincial consumer organizations (interview with M. Vallee, FNACQ, 1997).

The CRTC decided to grant Unitel and BCRL's application to enter the public long-distance voice-service market. The commission concluded that competition would provide benefits to consumers by lowering toll rates and increasing the choice of products. According to the CRTC this was very important for improving the international competitiveness of Canadian businesses. The CRTC, stated that competition would not jeopardize universal access to local services as long as the new entrants made adequate contribution payments to the Stentor members' companies to cover access costs, among other things (CRTC, Consumer Friendly Competition: The Facts, 1992: 1). Not only was the entry of Unitel and BCRL into the public long-distance market approved; the CRTC permitted total open competition, so that resellers could also compete in the long-distance market. The CRTC stated that long-distance telephone rates would now be able to drop, which would have particular interest for the heavy long-distance users (CRTC 92–12, 1992: 10–14). The commission further stated that it did not believe that a single-supplier environment (telephone monopoly) would be able to meet the needs and demands of the multiple users (59).

With regard to the issue of neo-regulation, the CRTC stated that regulatory constraints should not "be so burdensome so as to mitigate the beneficial effects of competition" (132). According to the commission, regulatory standards were put in place to prevent anti-competitive practices. This meant that the Stentor member companies would still be subject to tariff filings and rate-of-return regulation (132–3). In what would be one of many unusual temporary agreements among the competitors such as Unitel, the consumer and public-interest groups CAC and NAPO all agreed that the Stentor member companies should be subject to more stringent regulation.

Regulation for the competitors and the regulators, however, was quite a different matter. Unitel and BCRL, were subject to far less regulation. They could set telephone rates they considered appropriate as long as they submitted them in advance to the CRTC, which would do no more than ensure that the rates for the services offered complied with the Railway Act, whereas resellers would rely totally on market regulation and be able to enter and exit the long-distance market at will (141). According to the current CRTC vice-chairman, David Colville, this decision was the first of a number of liberalization and neo-regulation decisions in which the commission would oversee a "managed transition to a competitive environment" (interview with David Colville, 1996).

Most CRTC telecommunications decisions are arrived at based on a consensus among the commissioners after they take into consideration submissions, interrogatories, and the cross-examination of participants in the public hearings, and after a review and analysis is completed by CRTC staff. Occasionally a commissioner may informally record his or her dissent with the other commissioners in the final decision (D. Colville, 1996). Even more infrequently, a commissioner may make a formal objection and attach it to the original decision. A dissenting statement will not change the outcome of a decision once it is made, but the decision changes from a unanimous to a majority decision and becomes part of the decision record.

Commissioner Edward Ross issued a formal dissent statement for Telecom Decision 92–12. Ross was concerned, as had been some of the other commissioners, that local telephone rates would go up once competition was introduced to long-distance telephone service (D. Colville, 1996/2001). Ross's dissent stated that "no absolute assurance can be given that rates for local service will not be impacted by allowing this competition [long-distance telephone services]. Because of the subsidy provided to local services by long distance services, in allowing competition one can foresee millions of ordinary Canadians paying for the cost of competition through higher rates for local service" (205). Ross went on to explain that he thought that the public interest would best be served by allowing the single supplier (Stentor) reference plans. Commissioner Ross explained: "In conclusion, I reiterate that I consider the cost of allowing these applicants to be too high because of the increase in the cost of basic telephone service, something that is a necessity not a luxury for Canadians, including the millions of Canadians living on pensions and fixed incomes. We have been well served to date and the proposals put forward by the Stentor members would see further improvements to assist all users, including business" (206). Some have criticized Ross's dissenting decision as support for the status quo, for maintaining the monopolies of the incumbent telephone companies. But his comments reveal that he was very concerned that the cost of competition would be high, resulting in major increases to the basic telephone rates that many Canadians now considered to be an essential service.

INTENDED AND UNINTENDED POLICY CONSEQUENCES

One of the Stentor affiliates, Bell Canada, challenged the contribution portion of the long-distance-competition decision at the Federal Court of Appeal. Bell, acting now on behalf of the six federally regulated telephone companies, appealed the decision on two grounds. First, Bell claimed that the CRTC exceeded its jurisdiction in ruling that 70 per cent of the competition start-up costs should be borne by the telephone companies. Second, Bell argued that the CRTC also exceeded its jurisdiction in ruling that the contribution discounts, to keep local rates low, were far too generous. Despite Bell's complaint, the appeal was denied (Federal Court of Appeal, File A-900–92, 22 July 1992).

Bell subsequently responded by applying to the CRTC to increase its local rates[5] and reduce its local calling areas (Bell Canada, 1993). Although the CRTC initially denied this rate increase, it approved one year later what became a politically explosive two-dollar-a-month increase. Bell responded by taking a very pro-competition position, as outlined in its report *Freedom to Compete: Reforming the Telecommunications Regulatory System* (1993).[6] In this seminal report Bell publicly embraced competition and stated that now that the telecommunications market structure has been reorganized, a corresponding change was required for the regulation of the established telephone companies. The report advocated "focused regulation," where regulation for all telecom providers, not just the new players, would be reduced as much as possible (Shultz and Janisch, 1993: 7). The report further stated that the existing regulatory environment needed to be loosened to give the telephone companies freedom to compete for the acceptance of shareholders rather than subscribers (ibid., 8). Bell's supposed altruistic concern for the principles of universality was short-lived in the new competitive milieu that favoured shareholders at the expense of subscribers.

As Bell noted, the CRTC long-distance-competition decision ushered in a fundamental change to telecommunications market structure, from the structure of monopoly to that of oligopoly.[7] In an oligopoly, one firm can control 40 per cent to less than 100 per cent of the market, and the other leading firms along with the dominant firm may control from 60 per cent to up to 100 per cent of the market (Trebling, 1995: 1). By 1997, after five years of competition, a tight oligopoly market structure had unfolded in Canada, resulting in the incumbent companies retaining 67 per cent of the market share. The alternative long-distance suppliers AT&T Canada (formally Unitel) and Sprint Canada (formally Call-Net) held 10 per cent of the market, while Fonorola's share was 4 per cent (Industry Canada, 1999: 22).[8] Although there was an influx of long-distance resellers, from eighty in 1993 to two hundred in 1994, by 1996 only five resellers had significant market share. They were ACC Long Distance Corp., Westel, London Telephone, Distribute Tel., and Cam Net, who together accounted for 7 per cent of the long-distance telephone market (see Figure 6.2) (comments of AT&T Canada, Telecom Public Notice CRTC 96–26: 16, 1997; Industry Canada, 1996: 14). Despite the increase in the number of long-distance suppliers, with increased consumer choice and competition there is also more potential for undesirable business practices. The main issues raised by consumers have been aggressive marketing tactics, invasion of privacy, and slamming. Slamming occurs when telephone customers find out that they have been switched to another long-distance service provider without their consent.

The CRTC recorded 5,000 complaints and queries about slamming at its peak in 1995. These complaints subsequently dropped off to 3,000 in 1996 and 2,000 by 1997 (PIAC, 1998: 73). Total telecommunications complaints to the Telecommunication Ombudsman[9] were broken down thus: slamming, 69 per cent; disputed charges, 4 per cent; marketing, 15 per cent; billing, 2 per cent; failure to terminate, 2 per cent; telemarketing practices, 6 per cent; and other, 2 per cent (ibid., 74). In some cases customers had to spend considerable time negotiating outstanding bills before they returned to the provider they chose.

Telecommunications continentalization has also intensified as foreign players such as AT&T and Sprint U.S. increased their investment activities in Canadian telecommunications. For Unitel, a transfer of ownership was concluded in January 1996 that saw the corporation

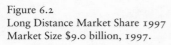

Figure 6.2
Long Distance Market Share 1997
Market Size $9.0 billion, 1997.

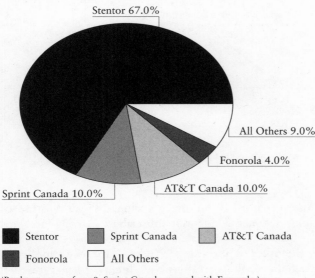

Stentor 67.0%

All Others 9.0%

Fonorola 4.0%

Sprint Canada 10.0%

AT&T Canada 10.0%

| ■ | Stentor | ■ | Sprint Canada | □ | AT&T Canada |
| ■ | Fonorola | □ | All Others | | |

(By the summer of 1998, Sprint Canada merged with Fonorola.)
Source: Adapted from Industry Canada 1999: 22.

change its financial and corporate structure as well as undergo a name-change to AT&T Canada. Under the new structure 33 per cent of the voting shares of the holding company, Unitel Communications Holding Inc., are now held by its subsidiary AT&T Canada. The other shares are held by the Bank of Nova Scotia with 28 per cent, Toronto Dominion with 23 per cent, and the Royal Bank with 16 per cent (Industry Canada, 1996: 14). Although share ownership of AT&T Canada is held by domestic and foreign multinationals involved in banking and telecommunications, corporate decision-making and management essentially rest with AT&T Canada (see Figure 6.3). As a private Halifax-based company, AT&T Canada is not obliged under provincial corporate law to issue public information such as annual reports other than what it posts on the Internet, or on its home page on the World Wide Web (interview with Peter Chaves, AT&T Canada, 1997). In March 1999 AT&T Canada merged with Calgary's Metro-Net.[10] As a result, AT&T Canada calls itself a "super carrier" – a one-stop source of continent-wide business telephone services. Interestingly enough, the company sold its residential long-distance business and residential Internet customers in order to focus on its corporate clients.

Figure 6.3
AT&T Canada share ownership as of 1996

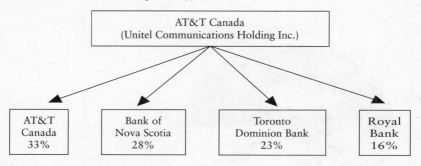

Source: Adapted from Industry Canada, 1996: 14

This raises serious questions about the growing unhealthy state of competition for long-distance residential subscribers in some areas of the country.

Sprint Canada, the other major alternative long-distance supplier, is 100 per cent owned by Call Net Enterprises Inc., a former reseller. Call Net also has 19 per cent ownership in Micro Cell Telecommunications, with a national licence to provide digital wireless telecommunications services. Sprint Communications Corporation (Sprint U.S.) holds a 25 per cent equity in Call Net Enterprises (see Figure 6.4). This has provided Sprint Canada with Sprint U.S. technology as well as direction and decision-making that come from three executives from the U.S. corporation, who also sit on the board of directors of Call-Net Enterprises. Through numerous acquisitions from 1993 to 1996, Sprint Canada now has a national and continental network that includes links in major cities from British Columbia to Newfoundland, with interconnections to Seattle, Detroit, Buffalo, Vermont, and Massachusetts (Annual Report, 1995, 1996).

Bell Canada has also begun investing outside Canada by purchasing a 5 per cent interest in MCI (Crandall and Waverman, 1995: 9). Since liberalization and foreign-ownership rule changes, continental integration among the major telecommunications and communication companies has increased. In January of 1990 BC Telecom Inc. and Telus Corp. (formerly Alberta Government Telephone Ltd.) merged into BC Telus Communication. BC Telus is able to benefit from technology and direction from its former parent corporation, GTE Corp., and Bell Atlantic, which has taken over GTE. BC Telus has subsequently taken on a more aggressive management style, announcing that it intended

Figure 6.4
Sprint Canada share ownership as of 1995

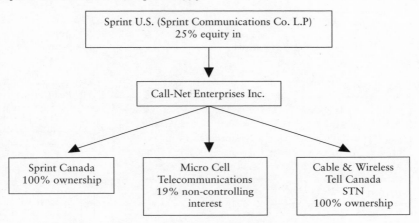

Source: Adapted from Call-Net Enterprises Annual Report, 1995: 1

to compete with Bell on its traditional turf in Ontario and Quebec as well as with the telephone companies in the Atlantic region. In the past the Stentor alliance prevented the telephone companies from competing in each other's territories. But once BC Telus announced that it intended to break this alliance, Bell/BCE responded by dissolving the organization in December 1998. Bell/BCE then went on the offensive, pushing west, east, and south. First, Bell formed strategic alliances with Manitoba Telecom Service Inc. and SaskTel. In the east the four telephone companies, Bruncor Inc. (NB Tel), Maritime Tele and Tel, Island Telephone Co., and NewTel Enterprises, merged into Aliant Corporation in the spring of 1999. Bell then increased its controlling interest in that corporation to 42 per cent. One day later Bell/BCE announced that Ameritech (a U.S. regional Bell operating company) had purchased a 20 per cent share of Bell. Figure 6.5 shows what Bell looks like after the reorganization was approved by the CRTC and the Competition Bureau. Included is BCE's 41 per cent share in NorTel Networks, its former telecommunications manufacturing company that competes with American and other firms continentally and internationally. Bell/BCE is, for the time being, part of the larger group of continental communications firms.[11] These mergers and alliances permit the development of continental oligopolies that can operate not only in the North American market but globally as well. Further, Ameritech gets a toehold in Canada, and Bell/BCE gains economies of scale and further access to U.S. markets.

Figure 6.5
Bell Canada and Amerirech network relations 1999

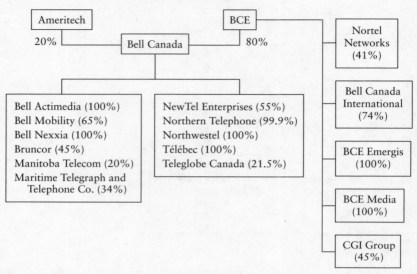

Source: Adapted with permission from Bell Canada Enterprises, 25 Mar. 1999.

CONTINUED CONSUMER AND PUBLIC-INTEREST ACTIVISM

After CRTC decision 92–12, which allowed any new telecom service company to enter to compete in the long-distance market, the commission accelerated the liberalization process and the move to neo-regulation. Neo-regulation was further extended by the CRTC through regulatory policies that split the base rate and prepared for price-cap regulation, permitting a continental regulatory environment that would be compatible with that of the United States. In a split-rate base the utility (access and local service) portion of the incumbent telephone companies is separated from the competitive side (long-distance services). The utility side continued to be regulated under the rate-of-return method until 1998. After that date, price-cap regulation took over. Until 1998 the rates charged by the incumbent telecom carriers were set so that the companies earned a rate of return on their common equity. As of 1 January 1998, any increases to the rates charged by the major incumbent telecom carriers for access to the PSTN have been determined by a formula price-cap index (PCI) that accounts for general price inflation and productivity improvements in the provision of telecommunications services.

The price cap sets the maximum percentage rate that incumbent telecom carriers can increase the prices they charge for a basket of capped services during a given period of time. Capped services encompass what are commonly referred to as the utility segment of the incumbent telecom carriers' operations. The utility segment is made up of basic residential local service, single and multi-line business services, and other services in which the telecom carriers still maintain an effective monopoly (e.g., operator service, unlisted telephone numbers, public telephone service, ISDN, digital-access service). Individual rate increases for residential and single-line business basic service are limited to no more than 10 per cent per year. There is also a Z-factor in the formula, representing exogenous factors that the CRTC may decide should affect the price-cap index. Examples include legislative, judicial, or administrative actions beyond the control of the company (Industry Canada, 1999: 19).

At the same time, the CRTC also freed the competitive side of the former Stentor companies from cost-based rate-of-return regulation, replacing it with a much lighter form of regulation through tariff filings and imputation tests. The commission mandated a carrier-access tariff (CAT), which applies to both the Stentor member companies and the interexchange competitors (IXC). CAT is based on a per-minute charge on each call rather than on long-distance trunk-line connections to each of the long-distance telephone companies (CRTC 94–19: 133, 163). CAT was subsequently appealed by AT&T Canada and Sprint Canada, and denied by the CRTC.

The CRTC sent out a public notice late in 1992 to examine whether the existing regulatory framework should be modified in light of developments in the telecommunications industry. Consumer and public-interest groups made their submissions for the upcoming public hearing but were surprised by the final decision, which included the liberalization of local telephone service, resulting in dramatic rate increases for basic local telephone service.

In spite of CRTC Chairman Keith Spicer's statement at the press briefing for Decision 92–12 that long-distance telephone competition would not increase local telephone rates, in less than two years the commission mandated local rate increases (CRTC 94–19: 22; see Appendix XVI for a complete list of participants)[12]. The decision sparked a public outcry from consumers' organizations and public-interest advocates, as well as from the new long-distance competitors. The consumer and public-interest groups mobilized into a broader alliance, called People for Affordable Telephone Service (PATS).

In this decision the CRTC made a move to reduce the cross-subsidization of local telephone service. Shifting to a neo-liberal policy approach requires telecommunications-service markets to be open to competition, while at the same time prices charged for local services are supposed to be based on economic costs (CRTC 94–19: 32). But as noted by the commissioners, reducing the local subsidy is different from eliminating it (Colville, 2001; McKendry, 2000). There are many areas of the country where the CRTC will never be able to bring local rates to economic costs because they would become unaffordable. Put simply, basic telecom services will always need some subsidy. Concurrently, the CRTC approved a rate increase of two dollars a month for basic local services, to begin 1 January in each of the years 1995, 1996, and 1997 (CRTC 94–19: 22). The commission chairman, claimed "With the introduction of local competition and policies promoting open access between network suppliers we believe that customers will have much greater choice in selecting a service package to meet their specific communication requirements … We want to create a network of networks to serve a variety of different consumer demands at prices that are fair and reasonable" (CRTC, News Release, 16 September 1994). No one bothered to tell local customers that greater choice would also carry a heavy price tag of seventy-two dollars more per year, nor to explain how fair and reasonable this was – or that all residential subscribers would not be able to benefit from local competition.

The non-profit organization the Public Interest Advocacy Centre had been conducting alternative telecommunications research and representing clients such as NAPO, the Federation Nationale des Associations des Consonmateurs du Quebec (FNACQ), and other organizations at the public CRTC telecommunications hearings for a number of years. These organizations and PIAC formed People for Affordable Telephone Service/Coalition pour un Service Téléphonique Affordable (PATS/CSTA). PATS represented the largest coalition of consumer groups ever assembled in Canada, with sixty organizations representing more than two million members. It included national and provincial consumer organizations, women's organizations, anti-poverty organizations, welfare equal-rights groups, teacher and student federations, the Council of Canadians, rural Canadians, and the Canadian Legion (for a complete list of members, see Appendix 18; PIAC Hotwire, March 1995).

The 1996 PIAC report *Consumer Coalitions: Three Case Studies* explains that many groups with different interests have been able to form into political-action collectives that are based on shared values, have commonly agreed-upon objectives, and have developed a collective consciousness that recognizes their increasingly disadvantaged social position (6). The immediacy of the local telephone rate increase helped to form a collective that tended to transcend class as well as organizational barriers. As a social group PATS members shared one common grievance – that the cost of telecommunications competition was being passed on to ordinary residential telephone users, to those with fixed, stagnant, or declining incomes, those least able to bear the cost of such a major overhaul of Canadian telecommunications policy.

The glaring inequality arising out of the local telephone rate increases further politicized the various organizations to form PATS, where they could pool resources and strategically fight the erosion of affordable basic telephone service. PATS/CSTA subsequently launched a petition to the federal Cabinet to reverse the portion of the CRTC decision that increased local telephone rates. PATS argued that there were strong procedural arguments that supported its petition. The initial CRTC public notice had not indicated that rate rebalancing or any other rate issues would be dealt with at the hearings. Consequently, consumer and public-interest groups were not prepared, in their submissions, their evidence, or their arguments, to counter local phone rate increases (PIAC, 1996: 16). In addition PATS/CSTA lobbied key government officials, members of the government, and the opposition parties, briefing them on the PATS position. PATS members also faxed the minister of Industry and publicized its efforts in an organizational newsletter and through mass-media advertising (17).

Parallel efforts were made by the newly formed Competitive Tele-communications Association (CTA, see Appendix 19). Acting on behalf of the interexchange competitors such as Unitel (AT&T Canada), Sprint Canada, and ACC, CTA petitioned the federal Cabinet to reverse the rate-rebalancing portion of Telecom Decision 95–21 and the two-dollar per month rate increase of CRTC 94–19 (see Appendix 16). The CTA's petition explained that it parallelled the PATS petition and sought an appeal of the largest CRTC-ordered rate increase in Canadian history because it had occurred without a proper public hearing (CTA, 1994: 3).

Although both petitions dealt with the same issue, it is important to point out none the less that co-operation took place only in terms of sharing essential facts. Both coalitions maintained their independence, and what happened in this parallel effort and other subsequent ones with the new competitors, the Stentor companies, and other organizations, in what seem like unlikely coalition alliances, occurs only over single and shifting telecommunications issues.

One year later, on the recommendation of Minister of Industry John Manley, the federal Cabinet delivered its response to the PATS and CTA petitions through an order-in-council. The order granted the Stentor companies local rate increases of two dollars per month, which would take effect 1 January 1996 and 1997. Additionally, the rate-rebalancing review was declined (P.C. 1995–2196, 19 Dec.1995). There were major differences between the order-in-council and the CRTC decision, aside from what at first appeared to be a lower consumer cost of forty-eight dollars for the two years concerned. There were three parts to the CRTC rate-rebalance proposal in Decision 94–19. On the utility side, the local rate would go up by two dollars per month to reduce the subsidy requirement. The commission was of the view that the competitive tele-com service side would pay the local service side less and at the same time the carrier-access tariff (CAT) would go down, whereas on the competitive telecom service side of the business, the CAT is an expense item. To make the whole rate-rebalance exercise work the CRTC ordered the telephone companies to reduce their basic toll schedule by the same amount that the CAT was reduced. This would have meant that the low-use long-distance subscribers who did not qualify for the various discount packages that were offered would also be able to obtain some benefit from competition – that is, it was to be strictly used to lower the long-distance rates of the Stentor companies for the small users until price-cap regulation came into effect (CRTC, 94–19: 23). It was this last step that the telecom companies successfully appealed.

As PATS demonstrated in its research, the majority of Canadian res-idential Stentor customers, in any given month, received no savings as a result of long-distance competition. The average long-distance bill for 66 per cent of Bell Canada's residential subscribers was eleven dollars a month. Yet Bell's minimum level to qualify for a discount applied only to customers whose long-distance charges were fifteen dollars a month or more (PATS, 1996: 7). The order-in-council decision, however, stated that significant price reductions in the long-distance market had already occurred and further price reductions would compromise tele-communications investment in research and development and innova-tive telecommunications services (P.C. 1995–2196: 2).

The order-in-council not only gave the Stentor member companies a substantial local rate increase; it did so unconditionally. The direct political intervention of the minister of Industry gave the Stentor com-panies what PIAC Executive Director Michael Janigan called "the biggest corporate bailout in Canadian history, worth $1.6 billion, where ordinary Canadians would make up the declining profits of the telephone companies in a competitive environment" (interview with Michael Janigan, 1997).

According to PATS and Stentor Policy Inc., Stentor out-lobbied the consumer and competitive interests (interviews with Richard Cavanagh, Stentor Policy Inc., 1996; PIAC, 1996). The Stentor companies pre-sented a counter-petition to the federal cabinet, requesting that regula-tory rate rebalancing should simply consist of local rate increases with no accompanying decrease in the direct-distance-dialled (DDD) rates (long-distance rates). Stentor accused the CRTC of tinkering with prices set by the open market (PIAC, 1996: 18). The petition was followed up with a letter from Stentor Policy Inc. to the clerk of the Privy Council, with copies to the CRTC and the minister of Industry, in which Stentor threatened to put a hold on research-and-development spending, which in turn would affect thousands of jobs in the advanced knowledge-based sector of the Canadian economy[13] (Stentor, 1995: 2).

PATS recommended to its members that the only way most long-distance residential Stentor customers (with the exclusion of Sask Tel) would receive lower long-distance rates would be to switch to alternative suppliers such as ACC Long Distance, Cam-Net, City Dial, CTI Telecommunications, Distrubutel, Fonorola, Funday Cable Ltd., London Telephone, Rogers Network Services, Sprint Canada, Unitel (AT&T Canada), or Westel (PATS, 1996B: 2,9). Although the order-in-council implemented local rate increases, concurrently the order and the minister of Industry made a public concession to the consumers' organizations for a lifeline program to ensure that universal access would continue for all Canadian subscribers, regardless of where they lived (P.C. 1995–2196: 2). The CRTC, and the CAC however, did not share this view.

AFFORDABILITY, ACCESS, AND UNIVERSALITY

Liberalization policies for long-distance services put pressure on the existing so-called subsidization of local-access telephone service. Additionally, liberalization and competition for both toll and local services, as requested by the established and new telecommunications providers, meant that local rates would have to move closer to costs. For ordinary Canadian residential subscribers this meant that their basic local telephone rates were already increasing by forty-eight dollars, with more increases planned to take place before price-cap regulation was established in 1998.

As the CRTC stated in its review of the regulatory framework, under the new liberalized environment the provision of telecommunications services would include placing greater reliance on market forces (CRTC 94–19: 3, see Appendix 16). The commission noted that regulatory reform could not continue with the existing subsidy scheme to keep rates for local-access service low. The CRTC went further and stated that the current subsidy was much larger than it needed to be to keep basic local telephone service affordable. In the regulatory-review decision, and in a subsequent one on splitting the rate base, the CRTC stated that a targeted subsidy or lifeline was not necessary at the time in light of mandated rate rebalancing that was already in progress (CRTC 96–10, 1996: 1). At the time the CRTC held the view that a lifeline progam for Canada was not necessary despite the fact that other countries, such as Australia,

New Zealand, and the United States, had, as part of their telecom policy liberalization process, introduced lifeline progams to make sure that low-income and rural residential subscribers would not drop off the existing telephone system (see Table 6.1).

Late in 1995 AGT and the Stentor member companies (excluding Sask Tel) submitted to the CRTC proposals that would essentially redefine basic local service. The CRTC responded by issuing a notice to hold a public hearing on local-service pricing options. In the public notice the commission stated that it would select a local-service pricing approach that would comply with subsection 7b of the Telecommunications Act 1993. This meant that the local-service approach selected would have to ensure that with the prospect of further local rate increases, particularly for residential subscribers, local services would remain universally accessible and affordable for all Canadians regardless of the region they lived in, whether urban or rural.

Two approaches were put forward, a budget-service model by the Stentor companies and a targeted-subsidy (lifeline) program, which was developed by the Fédérations nationales des associations de comsommateurs du Québec (FNACQ), the National Anti-Poverty Organization (NAPO), and One Voice (the Canadian seniors network), with support from other public-interest groups such as BCOAPO et al., the provincial CACs, and the Consumers' Association of Canada (CAC).

One very important issue dealt with at the public hearing was whether telephone service in Canada was affordable and whether basic local telephone service would continue to be affordable in the future. Most parties, such as Stentor, AGT, the Director, CCTA, the new competitors, the governments of British Columbia and Saskatchewan, and the CAC, maintained that the nation-wide telephone penetration rate computed by Statistics Canada was the most appropriate and reliable indicator of affordability. Stentor pointed out that overall penetration rates had increased from 97.6 per cent in 1983 to 98.5 per cent in 1995. Moreover, penetration rates for Canadians earning $10,000 or less was 93.9 per cent and for those earning $10,001 to $14,999, 97.6 per cent as of 1994 (Statistics Canada, 1994: 46–7). Stentor also provided evidence to show that out of 11.2 million households, only 169,000 did not have telephones. Of the 169,000 that had no phones, approximately 60,000 gave up their service because it was too expensive. The other 65 per cent were without a telephone because they either were in the process of moving or did not want a telephone (Stentor submission, 1996: 12).

FNACQ argued that although penetration rates are helpful, they are limited as an indicator because most people consider the telephone to be an essential service. People on fixed or with low incomes faced with

Table 6.1
Summary of Lifeline programs in Australia, New Zealand,
and the United States as of 1994

	Australia	New Zealand	U.S.
Flat rates	No	No	Yes[1]
Budget	Yes (easy-pay trial)	No	Yes[2]
Lifeline	Yes	Yes	Yes[3]
Federal targeted subsidy available	Yes	Yes	Yes[4]
Means tested	Yes[5]	Yes	Yes[6]
Administration	Telecom (Telstra)	NZ Federation of Family Budgeting Services	NECA[7]
Source of funding	Telecom (Telstra)	Telecom	IX[8] *carriers*

Source: Response to Interrogatory SRCI (CRTC), 4Apr.96–701, attachment 1, p. 1

1 Except Winsconsin, 1994.
2 Except Maine, 1994.
3 Except: Arkansas, California, Florida, Georgia, Indiana, Kansas, Louisiana, Maine, Michigan, Mississippi, Nebraska, New Hampshire, New Jersey, North Carolina, Oklahoma, Pennsylvania, Rhode Island, South Carolina, South Dakota, Tennessee, Texas, Vermont, 1994.
4 Except Delaware, 1994.
5 Contact (soft dialtone) has no means test.
6 Except Delaware and California, 1994.
7 National Exchange Carrier Association.
8 With at least 0.05 percent of nationwide pre-subscribed lines.

difficult financial circumstances will give up other consumption items, even food and medication, in order to keep their phone service (FNACQ, M. Vallée, 1996; NAPO, L. Toupin, 1997). BCOAPO et al. and FNACQ noted that the Statistics Canada penetration-rate methodology is problematic because the survey is only conducted every two years. As well, it excludes Canadians who live on Indian reservations, those living in the Yukon and Northwest Territories, and people who live in rooming- and boarding-houses (hotels). Moreover, the sample size is very low, at approximately 35,000, or 0.001 per cent of the population. Evidence presented by the Manitoba Keewatinowi Okemakanak (MKO) communities, representing more than 29,000 indigenous peoples living on twenty-six northern Manitoba First Nation Reserves, shows that household penetration rates of residential telephone service in the average MKO community is 69.1 per cent, compared with the Manitoba average of 98.3 per cent or the national average of 98.5 per cent. Moreover, penetration rates ranged from a high of 77 per cent in the community of Cross Lake to a low of 42 per cent in Shanattaawa (MKO, 1996: 1,

Table A.1). What FNACQ proposed was the use, in addition to penetration rates, of a benchmark of fifteen dollars, to be established as a reasonable indicator of affordability for low-income households (FNACQ et al., 1996).

The commission explained in its final decision that there was no consensus among the various submissions on a definition of affordable rates. The CRTC further stated that it did not believe that the previously approved two-dollar per month increase (for 1996, 1997, and 1998, which amounted to $72) would create an affordability problem. The only concessions the CRTC made were the extension of payments for installation charges, security deposits, and toll blocking. The CRTC did express some concern that forthcoming rate increases needed to be scrutinized through an "affordability monitoring plan." The plan was to be implemented by the telephone companies, the Stentor member companies, AGT, and Edmonton Telephone (CRTC 96–10, 1996: 8). The commission directed the telephone companies to file quarterly monitoring reports based on Statistics Canada's Labour Force Survey (LFS). The LFS statistics were to include penetration rates, non-subscription, and disconnection questions. This information was to be cross-filed with an annual supplementary report containing affordability analysis based on the socio-demographic statistics reported in Statistics Canada's HIFE microdata file. HIFE-based penetration rates were also requested at national and provincial levels, and by income (31–2).

Essentially, the affordability-monitoring plan puts the fox in charge of the hen-house. As part of the CRTC's vision of managed regulation, this elaborate technocratic scheme was favoured over the simpler low-income $15 benchmark. According to the CRTC, such a benchmark would still provide too broad a subsidy, one that was estimated to cover one million telephone subscribers rather than a more directed one of 200,000 (Rideout, 1996: 10). In the CRTC's view, current penetration levels revealed that at all income levels, local telephone services were still affordable. The commission also held the view that local telephone competition would lower basic rates, helping to keep them affordable (D. Colville, 1996). Yet the tracking by PIAC of residential telephone service rates reveals that the average basic rate increase in selected cities across Canada was 81 per cent as of 1998 (see Table 6.2). Table 6.2 shows that since the introduction of competition in long-distance phone service, rates in the large urban centres have almost doubled. In addition, rates in smaller cities and rural towns, particularly in Bell and BC Telus territories, have increased almost 120 per cent on average. This brings into question the claim by the telecommunications industry that we are witnessing the end of geography – and more importantly, that competition would bring down telecom rates.

Table 6.2

Selected basic residental-service rate increases 1992, 1997

City	Rates 1992 $	Rates 1997 $	Rates 1998 $	Increase $
BELL CANADA				
Toronto	13.70	21.30	21.60	58
Kingston	9.85	16.15	19.85	102
Arnprior	7.10	16.15	19.85	180
BC TELEPHONE				
Vancouver	14.90	21.25	24.10	62
Victoria	11.65	16.00	19.10	64
Prince George	9.85	16.00	19.10	62
TELUS (FORMALLY AGT)				
Calgary	12.18	20.90	22.00	81
Lethbridge	11.27	20.90	22.00	95
Bolye	10.47	20.90	22.00	121
MANITOBA TELEPHONE SYSTEM				
Winnipeg	13.70	17.30	17.65	29
NB TELEPHONE				
Fredericton	12.85	20.00	20.00	56
MARITIME TELEGRAPH & TELEPHONE				
Halifax	15.15	23.00	25.00	65
ISLAND TELEPHONE				
Charlottetown	13.60	20.20	22.25	64
NEWTEL				
St Johns	12.75	17.45	19.96	57
Total rate increase				81

Source: PIAC *Hotwire*, Sept. 1998. Adapted with permission.

A special CRTC hearing on local telephone service held in the fall of 1998 to deal with local rate increases for subscribers who live in rural and remote regions demonstrates that geography and distance still matter. The Telephone Service to High-Cost Serving Areas decision indicates that, as of 1999, the rural and remote areas of Canada had not benefited from competition. Consequently, the CRTC set out three goals to improve the situation for these areas: extending service to unserved areas; upgrading basic services to rural and remote areas; and finally, maintaining service levels to these areas to make sure that the existing base service does not erode under competition (CRTC 99–16, 1999). The commission has backed up these goals by continuing the "obligation to serve" for incumbent local-exchange carriers. This means that construction-service expenses such as upgrades and improvements should not be charged to the subscriber requesting the upgraded services. To ensure that the local-exchange carriers comply, the CRTC established a monitoring program to track service improvements. Another important aspect of the decision on service to high-cost areas is the announcement that another hearing will take place that will provide an opportunity for the incumbent local-exchange carriers to redefine their costing bands. The rationale for this hearing is to define the bands more narrowly and identify the high-cost areas in order to set up a narrow subsidy program for the high-cost areas. Again, part of what is taking place in the new competitive telecommunications environment is a narrowing or targeting of subsidies for local basic telecom service.

Another report, *Still a Long Distance To Go*, notes that before long-distance competition, the consumer price index (CPI) from 1984 to 1994 showed a general decrease in prices for long-distance services. The CPI from 1988 to 1991 fell at a faster rate than it did after competition was introduced. This disaggregated consumer price index, which was calculated by Statistics Canada, was, unfortunately, discontinued just as competition was established in the long-distance market (PIAC, 1998: 49). This suggests that there are politics in data gathering as well.

At the hearing to consider local-service pricing options, the CRTC was looking for a solution should local rates become non-affordable after the shift to cost-based pricing. The incumbent telephone companies (formerly the Stentor member companies) and AGT on its own proposed to offer discounts for local subscribers below the standard basic rate. The companies intended to do this through a number of budget optional-service packages. At first glance the companies' offer seemed magnanimous; but further examination shows that the companies intended to offer their customers lower rates in exchange for a reduced level of service. Depending on the particular company, offers were made for different forms of local measured service that included limits on free calling, reduced local calling areas, and restrictions on access and/or pay-per-use access to long-distance networks, all of which amounts to a two-tiered telephone service (CRTC 96–10: 9).

A coalition of consumer and public-interest groups, including FNACQ, NAPO, and One Voice (FNACQ et al.), countered that a targeted (lifeline) subsidy program should be put in place. This proposal suggested that basic local service be made available at a reduced rate to low-income subscribers. The proposal received support not only from other consumer and public-interest groups but from indigenous peoples, the large user organizations, and the long-distance competitors.[14] Although it would appear that the new long-distance competitors and FNACQ et al. made strange bedfellows, the new competitors supported the subsidy progam because they would be able to access the universal connection fund.

FNACQ et al. proposed that a universal-connection fund be established to provide a subsidy for low-income customers. Low income would be defined by Statistics Canada's LICO (low-income cut-off), and subscribership would be based on a self-certification basis to avoid the humiliation of any means-testing, similar to the California subsidy progam. Eligible subscribers would receive a flat local-service rate, and access to long-distance services would be capped at a rate of fifteen dollars a month. Similarly, installation charges would also be capped, but at twenty-five dollars a month. All telecommunications providers would contribute to the universal-connection fund through a revenue surcharge on all telecommunications services. An independent administrative body would oversee the operation of the progam (FNACQ/NAPO/One Voice, 1996). FNACQ et al. also included as part of its evidence a detailed report from D.A. Ford and Associates on the estimated cost to Canadians for the targeted subsidy plan. Other evidence included a survey of consumer perceptions about Canadian telephone service, commissioned by a broader coalition among labour, public-interest advocates, and consumer groups.[15] This survey countered the surveys conducted for AGT and Stentor because respondents stated that they considered the telephone an essential service; that Canadians were very concerned about telephone affordability; and that low-income subscribers should be subsidized (Ekos Research Associates, 1996).

It was expected that the Stentor member companies as well as AGT, Edmonton Telephone, Télébec, and the Ontario Telephone Association (independent telephone companies) would not directly support the targeted-subsidy plan. But an unexpected set of events occurred at the public hearing when the Consumers' Association of Canada (CAC) once again broke ranks with the other consumer and public-interest groups. CAC was of the view that the FNACQ et al. proposal was cast too broadly and was not necessary at the time. According to CAC, the existing Canadian telephone penetration rates in 1996 demonstrated that local affordability was not a problem. CAC suggested that one way to narrow subsidy eligibility would be to use welfare levels, which incidentally require means testing. NAPO countered CAC's welfare proposal by explaining that welfare schemes would pose a serious problem because most recipients receive benefits for only two years; only 25 per cent get jobs; and welfare progams do not address the low-income status of the working poor. In addition, many provinces do not include an explicit amount in a total social-assistance financial-aid package to pay for telephone service (Rideout, 1996: 10; Appendix 6). In fact NAPO, One Voice, and FNACQ felt betrayed by CAC. Although CAC shared in the expenses for the targeted-subsidy submission and survey, in its final submission to the CRTC, much to the surprise and amazement of the other consumer and public-interest groups, CAC publicly withdrew its support for a Canadian lifeline program.

The CRTC rejected the Stentor and AGT proposals and agreed that a targeted-subsidy progam would be the best way to address local residential-telephone affordability should a serious affordability problem arise in Canada in the future. But instead of implementing a lifeline progam, the CRTC mandated the established telephone companies to set up the monitoring progam. If an affordability problem developed in the future, the CRTC stated, it would develop a narrow-based subsidy progam, although no details have been provided of what the criteria would be to trigger the subsidy, or what the CRTC would use as a targeted-subsidy benchmark. In its final decision the CRTC agreed with the CAC that the low-income targeted subsidy, which was aimed at two million households, had been cast too wide (CRTC 96–10; D.A. Ford, 1996). One may also surmise that the incumbent telecom companies would not have been pleased about the subsidies they would have to contribute to a fund that would require $745 million in the first five years.

Once again the Consumers' Association of Canada helped to thwart a lifeline program for low-income Canadians. Other affordability mechanisms, supported by both the established telephone companies and all the consumer and public-interest groups, were approved by the CRTC. Other obstacles for low-income subscribers to accessing and remaining on the telephone network were addressed by allowing a six-month payment plan for installation charges, a three-month plan for security deposits, and free bill-management tools (CRTC, 96–10, 1996: 19–24).

The findings of a comprehensive internal review of the CAC, "Representing the Consumer Interest into the Next Millennium," sheds some light on why there was, and continues to be, such a major division in the consumer organizations over policy issues, particularly competition-policy issues (Intersect Alliance, 1998). This report identifies a few of the major problems that have plagued the organization, including a national and provincial policy and funding issues, poor national management of financial affairs, and the fact that the CAC is not the authoritative voice of the consumer perspective.

First, significant disagreements between the provincial and national chapters resulted in exclusive policy development and policy positions reflecting the views of only a vocal few. Consequently, the CAC's critical link through the provincial associations to consumers from across Canada was jeopardized. This would probably account for the differences between NAPO, FNAC, One Voice, Rural Dignity, and BCOAPO et al. and the national CAC on telecommunications competition-policy issues.

Second, in the past the CAC has had a spotty record managing its financial affairs, from its venture in the white-elephant purchase, sale, and loss of a national office in the Ottawa region to the failed *Canadian Consumer* magazine. Recent financial problems reveal that the CAC has been unable to adjust to core funding cuts by the federal government to conduct research on consumer issues. In the late 1980s and early 1990s the CAC received the lion's share of consumer development and project contributions, which peaked at $800,000 in 1991– 92. After telecom competition was in place, government development and project awards for the CAC continued to decline. By 1999–2000, project awards for the CAC were at $238,000 and more equitably distributed with other consumer organizations. This may just be a coincidence. Certainly a number of factors, such as competing consumer and public-interest credible research, would have impacted project awards. In addition, by 1996 and 1997 the CAC showed annual deficits of $100,000. Subsequently, a couple of showdowns occurred over major cash-flow problems. A temporary reprieve was granted with the establishment of the CAC, which allowed the private sector to make contributions to the organization. The first contribution came from the Canadian Bankers Association ($100,000), with others from Stentor ($30,000), the Canadian Cable Association ($15,000), and the insurance industry ($15,000), for a total of $160,000 (interview with A. Reddick, 1999). The establishment of the CAC Foundation raises serious questions about the organization's independence regarding consumer issues and policy positions for telecommunications, as well as banking and insurance. When the second cash-flow situation developed, the CAC fired its executive director and all staff except one, and continued to "hang on by its fingernails" with the help of member-volunteers. This band-aid solution to a serious organizational problem is likely to contribute to a continuation of the CAC's uneven research performance and policy positions.

This leads to the assessment report's third point, that "the extent to which the CAC actually speaks for a consumer constituency is a topic of major concern to the very government departments it seeks to influence" (Intersect Alliance, 1998: 13). The CAC no longer speaks as a grassroots movement representing the 170,000 members of its heyday in the 1950s and 1960s. Membership is down to 6,000, and a survey conducted by the CAC provides interesting demographic information. For example, of 772 member respondents, 90 per cent were over the age of 50; 65 per cent were over 65; and only 1 per cent under 35 (ibid., 41). According to some, provincial member-volunteers are drawn from this age-group and tend to come from the middle- and upper-middle-classes with little understanding of all consumer-policy issues.

Both the CAC and the Consumer Federation of America naïvely opposed the continuation of monopoly telephone service in Canada and the U.S. respectively. One of the major differences between the two organizations was that the CAC, as a matter of principle, supported competition, whereas the CFA opposed AT&T's telephone monopoly. In the U.S. consumer advocate Ralph Nader initially also supported competition. However, Nader later admitted that he had been wrong because competition did not end up serving the public interest; prices increased and quality of service deteriorated. The CAC, CFA, and Nader were naïve about the telecom sector, the market power of the incumbent companies, the natural trend of an unregulated market to dominance by one or a few companies, and in asuming that there would not be price increases in the absence of regulation.

In the U.S., the joint 2002 report by the Consumers Union and the CFA, *The Telecommunications Act: Consumers Still Waiting for Better Phone and Cable Services on the Sixth Anniversary of National Law,* questions the so-called competitive telecom environment in the United States. The report notes that technically long-distance rates dropped by 14 per cent. However, when long-distance companies lowered their per-minute rates at the same time, monthly fees increased. In addition, there are additional charges for universal service fees, as well as in-state service fees in certain states. As a result many consumers are paying more, not less for long-distance services. And competition in local phone service has shrunk as a result of the massive consolidation of two of the largest companies, Verizon and SBC, who now control 30 to 40 per cent of the country's local phone business. The report notes that consumers pay approximately two dollars a month more for each basic phone line and three dollars more for a second line (2000: 3). The major difference between the CFA and the Canadian Consumer organizations and the public-interest groups is that the latter were not rigorous enough in tracking or responding to the impact of telecommunications service competition.

In addition to basic telephone affordability and other major telecommunications policy changes, the effect of policy liberalization on local basic residential subscribers redefines, narrows, and reduces telecommunications affordability and universality. Affordability is redefined as fair and reasonable, but with the existing basic local-rate increases and planned increases, it is neither. In addition, the elimination of the ostensible subsidy to local telephone service has adversely affected subscribers with low and fixed incomes. Moving local rates closer to cost has resulted in significant CPI rate increases for residential subscribers, affecting very low-income subscribers and those subscribers who live in rural or remote regions of Canada, in addition to other increases. Until some kind of subsidy is put in place for low-income and rural/remote subscribers, we will continue to see advancing in inequality and a widening gap between the telecommunications haves and have-nots. Already there is evidence that as Canada moves towards establishing an information/knowledge economy, the gap is widening between those households that have information and communication technologies and access to the Internet and those that do not (Ekos Research Associates Inc., 2000; Dickinson and Sciadas, 1999; Reddick, 2000). Access, then, becomes a very important issue because the first level of connection will likely be through the telecommunications (penetration 98 per cent) or the cable (penetration 75 per cent) networks.

Neo-liberal telecommunication-policy changes identify the past objectives of affordability and universality as too broad. The aim has been to narrow the existing universal progam and to target subsidies to those who live in high-cost areas or those who can show that they really are poor. Unlike other countries, Canada has decided to take a wait-and-see approach before implementing a telephone targeted-subsidy program for low-income people. Such an approach invites the question: how many people will have to lose their telephone service before the federal government is willing to introduce some form of subsidization to those most adversely affected by its neo-liberal policy changes? Already those who cannot afford a phone and whose only connection to Canadian society is through a very low-cost voice-mail service have been forced to rely on the charity of groups such as the United Way and the benevolence of the private sector (*Toronto Star*, 15 May 1997: B6).

It will be recalled that liberalization of local telephone services was accompanied by major increases in local basic residential services, and that the CRTC directed the telephone companies to track those residential subscribers who gave up their phone, whatever the reasons. The annual monitoring reports submitted by Stentor and Bell et al. to the CRTC show that residential telephone drop-off in Canada has doubled since 1995. These reports rely on Statistics Canada's Household Income and Facilities and Equipment survey data. To date, no independent analysis has been conducted on the Bell/Stentor reports. The Statistics Canada data from 1995 show that approximately 60,000 households could not afford basic telephone service (CRTC Interrogatory, 1998). It will be recalled that, at the local-service pricing-options hearing in 1998, Stentor informed the CRTC that telephone penetration rates were

Table 6.3
Canadian households that
could not afford telephone
service 1995–99

Year	Households
1995	65,000
1996	132,500
1997	123,431
1998	127,294
1999	126,183

Source: Bell/Stentor Quarterly
Monitoring Reports,
1995–2000. Adapted with
permission.

high for all Canadians, including lower-income subscribers. Stentor assured the CRTC that Canadians did not have a telephone-affordability problem; only a few households (60,000) could not afford basic phone service. According to Sentor, 33 per cent of those subscribers gave up their telephone when they moved; another 33 per cent did not want or need telephone service; and 34 per cent gave up their phones because of affordability problems. This meant, according to Stentor, that only 20,400 households gave up their phones because they could not afford the service. The monitoring reports reveal, however, that not only has the number of households without telephone service increased; affordability is cited as the major reason for drop-off (see Figure 6.6). There is also an increasing class aspect to not being able to afford a phone, particularly for those households headed by female lone parents and single males who rely on government transfer payments as a source of income (Stentor, 1998: 24, Appendix 1: 20).

Based on the Bell/Stentor monitoring reports, from 1995 to 1999 Canadian households that cannot afford telephone service have more than doubled, from 60,000 in 1995 to 126,183 in 1999 (see Table 6.3). If we make a conservative assumption that there are on average four people in each of these households, this means that there are almost a half-million people in Canada who cannot afford access to the telephone network. Moreover, the monitoring reports show that from 1997 to 1999, for those households with a family income below LICO (73 per cent), affordability was the major reason for not having a telephone. The two major affordability problems were installation charges (72 per cent) and basic service charges (66 per cent). Only

Figure 6.6
Reasons for terminating telephone service by province 1997 (%)
Telephone Service Termination, 1997

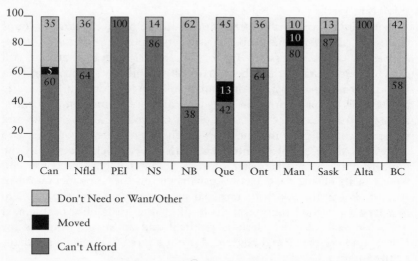

Source: Stentor Resource Centre Inc., 1998, App. 2, p. 5. Adapted with permission.

27 per cent of respondents cited moving/other as the major reason for not having phone service. Bell/Stentor's reliance on the HIFE data file has other methodological problems,including the low sample size of approximately 44,000 households and the fact that the Statistics Canada survey is only conducted every two years, as well as other sampling errors in gathering data from low-income subscribers. As Bell/Stentor and PIAC provincial analyses of low-income telephone penetration rates indicate, the rates are higher in Alberta, Saskatchewan, Manitoba, Ontario, and Quebec (averaging 95 per cent), and lower in British Columbia (92 per cent), New Brunswick (93 per cent), Prince Edward Island (93 per cent), Newfoundland (92 per cent), and Nova Scotia (89 per cent) (Bell et al., 1999; PIAC, 1998). At the hearing into local-pricing service options, a U.S. expert on lifeline programs, hired by Telus, informed the CRTC that any drop below 95 per cent in telephone rates should be of major concern. In addition, a 95 per cent level of confidence in the actual number of households without telephone service in Canada could vary as much as 20 per cent from the reported count of households. For policy-makers and regulators, a key factor in the issue of telephone access and affordability, as indicated in prior

research and surveys, is that Canadians consider the telephone to be an essential service, one that they are reluctant to give up, and will cut back on other essentials such as food, or even medication, to stay connected.

At the invitation of the CRTC, PIAC and the other consumer groups discussed the telephone drop-off issue. Its report *Eliminating Phonelessness in Canada: Possible Approaches* (2001) suggests that PIAC still does not understand that the CRTC will not likely be receptive to a broad low-income subsidy. Although a subsidy to address affordability problems is still a priority for consumer and public-interest advocates, according to Reddick, "PIAC only has a minimal amount of time and resources to track low income drop-off and we are the only group who represents these people" (interview with A. Reddick, 2000). In the hope that the Prime Minister's Office would be more receptive, on behalf of its clients PIAC has been working on the implementation of social policy in the form of a targeted subsidy to deal with the problem of telephone affordability and drop-off. The affordability problem, if not addressed, also has serious political and social implications for federal and the provincial government initiatives to connect Canadians to the Internet.[16]

7 Continentalizing Canadian Telecommunications

Canadians still look to their governments to insulate them from international economic forces, despite the fact that Canadian governments, including the federal government, have been emphasizing the need to adapt and adjust to market forces. Privatization, deregulation, the Free Trade Agreement, the Mexican trade initiative and reinforced attempts to achieve expanded General Agreement on Tariffs and Trade arrangements are all cases in point. As a result, many [Canadians] feel betrayed and bereft, and are confused and angry. Part of this is due to their sense that traditional Canadian values are being usurped by anonymous market forces and that governments are doing nothing to deal with these.

Canada, *Report to the People and Government of Canada*, Citizens' Forum on Canada's Future, 1991

We believe that our success in the past and our viability in the future are rooted in our ability to organize and build alliances with other unions, community organizations, consumer and public interest groups, and, in some cases, even governments.

Rod Hiebert, President of the Telecommunications Workers Union, "Building Alliances – The Key to the Future of the Labour Movement," address given to the Postal, Telegraph & Telephone International xiv Inter-American Congress, 1996

THE CANADIAN TELECOMMUNICATIONS ACT, 1993

In 1992 the Department of Communications presented a third telecommunications bill, C-62, to Parliament. Communications subcommittees were set up by the House of Commons and the Senate to examine the bill. These subcommittees served as the final semblance of pluralist input and participation for groups representing different social interests. But many of these groups were successful in influencing only minor wording changes in the final Telecommunications Act.

Bill C-62 was supported by the Stentor member companies, the alternative long-distance competitors, and the two major telecommunications-user organizations, the Canadian Business Telecommunications Association (CBTA) and the Information Technology Association of Canada (ITAC). ITAC supported the bill because it would consolidate and modernize the Railway Act. The bill also supported a more open domestic market, which in ITAC's view would improve telecommunications-equipment trade and increase the competitiveness of the telecommunications users

(ITAC, 1993b: 3). Similarly, the CBTA supported the bill because it helped to restructure the telecommunications environment so that it would be more favourable to a competitive marketplace. The CBTA was also pleased that the federal Cabinet (Governor in Council) and the minister overseeing telecommunications were given the power to override CRTC rulings (Canada, Commons, Bill C-62, section 8: 4). Concurrently, the CRTC would have more scope to design regulation to suit specific purposes and offer more regulatory flexibility (CBTA, *Issues*, 1–5).

Despite objections from two provincial governments, consumer organizations, public-interest advocates, and two of the largest telecommunications unions, Bill C-62 received royal assent in June 1993. In October of the same year the Telecommunications Act was added to the array of Canadian federal communications legislation. The objectives of the new act were written to reflect the economic changes that had occurred to Canadian telecommunications policy through the 1980s and 1990s. The following is an excerpt from Section 7 of the act, with a list of the key neo-liberal telecom objectives:

a) to facilitate the orderly development throughout Canada of a telecommunications system that serves to safeguard, enrich and strengthen the social and economic fabric of Canada and its regions;
b) to render reliable and affordable telecommunications services of high quality accessible to Canadians in both urban and rural areas in all regions of Canada;
c) to enhance the efficiency and competitiveness, at the national and international levels, of Canadian telecommunications;
d) to promote the ownership and control of the Canadian carriers by Canadians;
e) to promote the use of Canadian transmission facilities for telecommunications within Canada and between Canada and points outside Canada;
f) to foster increased reliance on market forces for the provision of telecommunications services and to ensure that the regulation, when required, is efficient and effective;
g) to stimulate research and development in Canada in the field of telecommunications and to encourage innovation in the provision of telecommunications services;
h) to respond to the economic and social requirements of users of telecommunications services;
i) to contribute to the protection of the privacy of persons. (Telecommunications Act, 1993, chap. 38: 1478)

The minister of Communications indicated that these objectives were much more than an innocuous updating of the Railway Act or mere telecommunications housekeeping (1993: 2–3). Essentially the act establishes an economic set of rights for the largest telecommunications users, namely the conglomerates and multinationals. Section 7 (f) in particular, which establishes reliance on market forces, is not so much a policy objective as a means to an end. Moreover, these economic rights are juxtapositioned with vague and watered-down social objectives in which the previous objective of affordability is replaced with reliable, affordable, high-quality accessibility. The universality objective, meant to address the government's concern that all Canadians should be guaranteed access to the telecommunications system, had been abrogated by a form of regionalism: people living in different regions, or in different urban or rural areas, could now be offered different levels of service and pay different rates rather than have access to an affordable common level of service. From a policy point of view Canadians would be going back again to the future, reverting to a telecommunications market model similar to the one that existed prior to 1905. Of the regulations introduced in 1905 to curb unfair market practices, very little was left in the way of regulation to protect residential subscribers.

These contradictory objectives have already increased conflicts, as the various policy players use the specific objective that best suits their position. A good example was Stentor's use of Section 7 (f), reliance on market forces, as the rationale for its budget-service offering as a way to address the increasing affordability problems resulting from liberalizing policies in local basic telephone service. By contrast, consumer and public-interest groups (FNACQ et al.) used Section 7 (b), the affordability objective, at the same public hearing on local measured service. The Department of Industry proudly refers to the Telecommunications Act as a new one that is pro-competitive (Industry Canada, 1999: 44). This contentious legislation exacerbates tensions and conflicts between economic and market forces on the one hand and social requirements, identity, and sovereignty on the other.

The Communications and Electrical Workers of Canada (CWC) and the Telecommunications Workers of Canada (TWU) were strong opponents of Bill C-62 because it would allow the federal government to use its jurisdictional powers and the new legislation to implement the pro-competition and neo-regulation policy agenda (CWC, 1993; Canada, Commons, 1993, Issue 4). The effect of these policies on negotiating would result in a major loss of telecommunications jobs. The governments of Saskatchewan and Quebec held similar views in their opposition to the bill because it would limit provincial input into any future national telecommunications policy (Canada, Commons, 1993a, 4: 43–57).

What was surprising was the position taken by the International Brotherhood of Electrical Workers (IBEW). The IBEW supported Bill C-62 because, in their view, the bill would enable the telecommunications industry to create a level regulated playing-field and remove all intercontinental and interprovincial barriers. The union's position was so surprising to the Senate Subcommittee on Transportation and Communications that one of the senators commented that the union sounded more like the pro-competition business lobby of the Chamber of Commerce (Canada, Senate, 1992, 17: 58). Internal divisions like this one within the Canadian telecommunications union movement have been continually exploited by the state. The general thrust of the federal government, whether it is involved in changing specific telecommunications policy or, in this case, introducing neo-liberal and neo-regulatory legislation, has been to take advantage of any divisions and fractionalization that appear within the major consumer and union organizations that continue to resist these changes.

The national umbrella union organization, the Canadian Labour Congress (CLC), countered the IBEW's position, as did the CWC and the TWU. All three were strongly opposed to Bill C-62 because it would eliminate or lessen the regulation of the telecommunications industry and because the bill helped, along with the North American Free Trade Agreement, to entrench legislation, policies, and a trade agreement that would further the continentalization of telecommunications. The CLC was concerned that reliance on market forces and competition would result in skyrocketing local rates, causing poor Canadians to drop off the telephone system. In addition, Canadians living in rural and remote areas would find access to the telecommunications systems unaffordable (Canada, Senate, 1992, 18: 18–62).

The CWC and the TWU were also concerned about the changes to the foreign-ownership and investment rules in Section 16 (3) (Telecommunications Act, 1993, chap. 38: 1483). Although the act continues to emphasize that Canadian ownership must be 80 per cent, clauses (b) and (c) provide ways around this. The new legislation makes concessions to what constitutes Canadian control of a carrier where only 80 per cent of the board of directors need be Canadian. Foreign ownership has also increased to 33.3 per cent, either by setting up a holding company or through direct or indirect ownership of a corporation's issued voting and outstanding shares (ibid., 1483). In addition, the inventive and creative ways corporate lawyers and accountants set up these holding companies effectively allow them to own up to 47 per cent of a carrier (interviews with Doug McEwan, Industry Canada Telecommunication Policy, 1997; Bob Tritt, Stentor, national director of International Affairs, 1997). No one is quite sure how the Canadian increase of foreign ownership to 46.7 per cent was arrived at during the World Trade Organization negotiations. It would appear on the face of it that the federal government negotiated something less than 50 per cent. In addition, the prospect of an electronic information highway where telecommunications carriers, broadcasters, cable-television operators, and the computer industry are converging, anything more than 50 per cent foreign ownership would unravel the foreign-ownership limits specified in the Canadian Broadcasting Act.

These foreign-ownership and management-control changes have furthered telecommunications continentalization. Both the CWC and the TWU were very concerned that these changes in ownership rules would result in enormous job losses in the telecommunications sector (Canada, Senate, 1992, 18: 61–2). In its testimony to the House of Commons subcommittee, the TWU provided evidence that job loss had already started to occur. At Telus (AGT) 1,200 jobs (IBEW workers) were eliminated; 600 jobs were lost at BC Telephone; and 70 jobs were lost at the Canadian Association of Communications and Allied Workers (Canada, Commons, Issue 4, 1993). The unions' concern about telecommunications job loss was well founded. Research conducted for the Department of Human Resources tried to present a positive picture: although there had been employment losses in the telecommunications service sector, there had also been job gains. The *Human Resources Study of the Canadian Telecommunications Industry* (1996) indicates that these job losses occurred in the clerical and communication-equipment installation and repair sectors, and among managers, telephone operators, and fabricating and assembly workers between 1991 and 1995, amounting to the elimination of 13,000 jobs. The report states that these losses have been offset by significant growth in the same period in sales and marketing, and among electrical and electronic engineers and computer hardware and software specialists (Department of Human Resources, 1996: 11; Interim Report, 1995: 51–2). This so-called job growth, however, only amounts to 2,800 jobs. The net loss amounts to approximately 10,000 workers. Figures compiled by Industry Canada paint a far bleaker picture. Over a six-year period the traditional wired telecommunications carriers lost a total of 35,433 jobs (Table 7.1). Jobs created as of 1995 by the alternative long-distance carriers (2,800) and the resellers (220) had amounted to only 3,020. The net loss in the Canadian telecommunications service sector amounted to 32,000 (Industry Canada, 1997: 8).

Industry Canada's report *The Canadian Telecommunications Service Industry* traces industry employment from 1984 through the period of policy liberalization and competition to 1998 (see Table 7.2). Employment levels were down 17 per cent from their peak in 1990, before the introduction of long-distance competition. With the exception of employment levels in 1995, since the introduction of competition,

Table 7.1
Canadian telecommunications employment figures 1990–95

Carriers Stentor	1990	1991	1992	1993	1994	1995
BC Tel	14997	15015	14524	13478	13797	13851
Bell Canada	56530	54632	52897	50982	51503	48333
Telus Corp (AGT)	10746	10201	9753	7739	7295	9539*
Sask Tel	4216	3981	3861	3699	3863	3845
MTS	4805	4739	4563	4408	4257	3956
NB Tel	2591	2432	2348	2283	NA	NA
MT&T	4300	4035	3899	3786	3703	3219
Island Tel	355	343	336	329	338	293
New Tel Ltd.	2141	2072	2037	1939	2128	2004
TOTAL STENTOR	100681	97450	94218	88643	86884	84726
TOTAL INDEPENDENT						
TELEPHONE COS.	6617	6529	6542	6541	6476	4343
ALTERNATIVE TELECOM CARRIERS						
AT&T Canada						1150**
Sprint Canada						1150
FONOROLA Inc.						500
TOTAL ALTERNATIVE						2800
CARRIERS						
TELECOM RESELLERS						
ACC Tel Ltd.			62	126		220
Smart Talk			NA	NA		NA
Cam-Net Inc.			NA	NA		NA
TOTAL ALTERNATIVE	0	0	62	126	0	3020
LONG DISTANCE PROVIDERS						

Source: Emplyment figures compiled by Industry Canada, 1997.

 * Includes Ed Tel.

** Estimated number of employees based on market share of 9 per cent, which is equivalent to Sprint Canada.

levels had continued to drop beyond the 1984 figures. Together Tables 7.1 and 7.2 support the telecom unions fears that employment in this sector would shrink (Industry Canada, 1999).

The Public Interest Advocacy Centre (PIAC) and the Fédération National des Associations de Comsommateurs du Québec (FNACQ), like the TWU and the CWC, were very concerned about Section 34 of Bill C-62. This section identifies another form of neo-regulation, forbearance. Essentially the forbearance section of the act gives the CRTC the power to refrain from regulating a telecommunications service providing it considers the market competitive enough (Telecommunications Act, 1993, Section 34). In its regulatory review the CRTC cited Section 34 as justification to forbear from regulating the sale, lease,

Table 7.2
Canadian telecommunications service industry
employment 1984–98

Year	Telecom service industry employment (persons)	Annual growth %
1984	112,700	–
1985	111,900	–0.7
1986	112,200	0.3
1987	114,400	2.0
1988	115,200	0.7
1989	124,700	8.3
1990	127,100	1.9
1991	125,000	–1.7
1992	121,700	–2.6
1993	107,500	–11.7
1994	110,600	2.9
1995	115,600	4.5
1996	104,100	–10.0
1997	103,100	–0.9
1998*	105,000	1.7

Source: Industry Canada, *The Canadian Telecommunications Service Indusry: An Overview 1997–1998*, 1999; and adapted from Statistics Canada, *Survey of Employment Payrolls and Hours*, Cat. 72–002.
*Average for the first nine months of 1998.

and maintenance of terminal equipment (Telecom Decision CRTC, 94–10, 1994). The Consumers' Association of Canada lent its support to the bill and was not concerned about the forbearance clause. Forbearance would, in the CAC's view, give the CRTC the flexibility it needed to regulate, or deregulate, the telecommunications industry as necessary (Canada, Senate, 1992, Issue 17: 25). PIAC, however, pointed out that forbearance would likely result in not enough attention to the role of regulation, and that regulation was just as necessary in a competitive environment. PIAC explained that regulation serves many purposes other than just economic ones, such as the achievement of social goals, among them universal availability and affordability of basic service (Canada, Commons, 1993a, b, 7: 9). Concern was also expressed by both PIAC and FNACQ that forbearance would give way to neo-regulation by the Director of Competition and the Competition Act.

According to CRTC Vice-Chairman Colville, forbearance is the key element in the act for the commission (D. Colville, 1996). Above all, forbearance is consistent with the overall neo-liberal approach taken by the CRTC to manage competition. If forbearance does not work, then, according to Colville, the CRTC can step in at any time and re-regulate. But as PIAC points out, there is nothing specific or intrinsic in the act about re-regulation. The way the act is written, public-interest and consumer groups would have to rely on the magnanimity of the CRTC to re-regulate.

Colville provides a useful neo-liberal map to explain how the commission is now proceeding. First, the CRTC introduces competitive policies. Then the commission must be flexible with the new entrants and keep a watchful eye on the dominant players. The next step involves price-cap regulation for telecommunications rates. Price-cap regulation, however, should not be confused with detariffing. Under price-cap regulation, the alternative to rate-of-return (RoR) regulation under monopoly conditions, there are still tariffs for services. According to the commission, price-cap regulation overcomes one of the major criticisms of RoR, which is the incentive to overexpand on capital. The CRTC can then forbear, in part or in whole. Once forbearance replaces regulation, then the Director of Competition can oversee the competitive telecommunications marketplace (interview with D. Colville, 1996, 2001). For example, Section 34(1) of the Telecommunications Act gives the CRTC the power to forbear from regulating a service or a class of services. Under Section 34(2) the act emphasizes that the CRTC must forbear if there is sufficient competition for the users. Section 34(3), however, directs that the commission must not forbear if a competitive market has not developed, or if a competitive telecom service market has eroded. The CRTC can also impose conditions on when forbearance can be granted. In many of its forbearance determinations the commission has still retained certain powers under the act in order to address anti-competitive behaviour or undue preference. Based on these detailed criteria, the CRTC has forborne in terminal equipment, long-distance telephone services, private-line services, data services, wireless services, inside wiring, satellite services, and international services, as well as the Internet (Industry Canada, 1999: 58–9). Regardless of how the transition to market regulation is managed by the CRTC, these neo-liberal policy moves still shift telecommunications services away from the public interest to market interests, eroding the social and utility aspect of telecommunications.

THE COMPETITION ACT AND
TELECOMMUNICATIONS LEGISLATION

As indicated by the Telecommunications Act and Commissioner Colville, neo-liberal regulation now requires more federal government structural management, not less. It includes the federal Cabinet, the minister responsible for telecommunications, the minster of Industry, the CRTC, the Telecommunications Act, the Director of Investigation and Research (the Director), and the Competition Act. The Director also plays a pivotal role on behalf of the federal state in advocating a neo-liberal telecommunications policy environment and a pro-competitive telecom marketplace.

The Director's mandate comes from the 1986 Competition Act. The act, as part of other larger federal initiatives such as the Macdonald Commission and the Canada–U.S. Free Trade Agreement, was the outcome of a major overhaul of the former Combines Investigation Act. According to the Director, competition is essential for an effective market economy. As such, the act encourages competition through businesses that are productive, enterprising, and efficient. For consumers, the act is supposed to provide competitive prices and more product choice (Director, 1993: 1–2). The Director enforces the act through the Bureau of Competition Policy. Until 1993 the Director and the bureau were part of the Department of Consumer and Corporate Affairs. This changed when Consumer and Corporate Affairs was folded into the Department of Industry, part of the massive federal government reorganization project to downsize the civil service. Within Industry Canada the Director sits as part of the management team, holding a position equivalent to a minister who takes part in policy discussions in the federal Cabinet (interview with Dr Gerald Robertson, co-ordinator, Regulatory Economics, Economics and International Affairs Branch, Competition Bureau, Feb. 1997).

The Director has shown a marked interest in telecommunications policy issues (Director of Investigation and Research, *Annual Report*, 1995). The Director, armed with the Competition Act, and the CRTC have been the most significant federal state agents to have helped to redefine telecommunications subscribers as users and consumers. Of particular importance is the attention the Director pays to the choice consumers now have in long-distance telephone service and to less regulation (Director of Investigation and Research, 1996: 6). The CRTC, by contrast, tends to redefine subscribers as users, and all policy outcomes "in the public interest" are considered favourable. When telephone subscribers are redefined as consumers/customers it helps, as

Mosco writes, the "commodification process at work in the society as a whole perpetuate communication processes and institutions, so that improvements and contradictions in the social commodification process influence communications as a social practice" (1996: 142). The social practices of the large telecommunication users have contributed to shifting telecommunications policy in a direction that emphasizes commodification. Commodification has, however, had a serious impact on all other residential users/consumers. Within residential-subscriber categories the focus on consumptive capabilities stratifies consumers into large-business consumers, upper and middle-class consumers, and lower-class or fixed-income consumers, the position supported by the Consumers' Association of Canada. This means that lower-income subscribers are not significant or vital enough in a telecommunications environment that is guided by the crassness of the market. Telecommunications commodification, with its emphasis on commercialism, encourages an individual-consumption philosophy, and is more conducive to continentalization in general. Recent research conducted by Henry Mintzberg warns governments of the dangers of viewing citizens only in turns of market and economic criteria (customers and consumers), which ignores the fact that people are citizens with rights and that government has obligations to them (Mintzberg, 1996: 76–8).

The success of the Competition Bureau and the Director in moving telecommunications towards the neo-liberal competitive policy model has further helped to unleash political actors who claim that the CRTC's role as a regulatory body is no longer necessary. This is the position that Bell Canada now takes, having whole-heartedly embraced competition after it lost the long-distance public hearing. The Competition Bureau accommodated this position by holding a "Telecom Antitrust Symposium" in the fall of 1995. The objective of the conference was to discuss the progress of regulatory reform within the framework of the broader Competition Act as opposed to sector-specific regulation by the CRTC. The conference brought together neo-liberal and pro-competition policy experts from Canada and the United States, provincial and federal bureaucrats, and representatives from the telecommunications industry.[1]

The alternative long-distance carriers, the CRTC, PIAC, FNACQ, CWC, and TWU do not share the view put forward by the Competition Bureau that the CRTC regulator had outlived its usefulness. Certainly, eliminating the CRTC would only benefit the dominant telecommunications companies of the Stentor group. For the unions, the consumers' organizations, and public-interest groups, ending CRTC regulation would end any pretence of pluralism. If the Competition Bureau were to replace the CRTC, telecommunications abuses would be investigated using only economic criteria, examining market dominance and anti-competitive behaviour. Complaints would be handled by one of the bureau branches that investigates and determines whether complaints are relevant. If they were, they would then go to the Competition Tribunal. This form of regulatory undersight would have the Director, the bureau, and the tribunal acting as investigator, judge, and jury. Moving to this form of market regulation would eliminate what is left of public participation in the telecommunications policy process, leaving residential subscribers with little if any recourse. Furthermore, such a move would foreground the Competition Bureau and the Competition Act over the CRTC.

How competition issues would be treated by the Director and the Bureau of Competition is already evident in the way it dealt with the Hollinger takeover of Southam Press. In 1996 Hollinger Inc. purchased the Southam Press Inc. newspaper chain. This takeover resulted in Hollinger's controlling over half of the daily newspapers in Canada and more than 40 per cent of the daily circulation (Mosco, 1997: 163–4). The Director of the Competition Bureau approved the deal, using economic criteria for its assessment. As a result of the Director's ruling, the case did not even go on to be heard at the Competition Tribunal. Using only economic criteria to assess the effects of this merger, social and cultural impacts dealing with content issues were totally ignored.

In August 1996 the federal government issued a convergence policy statement, establishing broad policy objectives for telecommunications and broadcasting in the context of the information highway. This statement was the outcome of a public consultation process that was initiated by an order-in-council to investigate the status of network facilities, resale and sharing, and the production and exhibition of Canadian broadcasting content. Technological advances in telecommunications, data communications (computer), and mass communications such as broadcasting have resulted in the structural convergence of these industries. Integration may put together sounds, data, text, and images, combine different transmission links and equipment, including telephone (voice) and computer (data) communications, or it may combine information and communication services on the Internet. Key here is the convergence of telecommunications, broadcasting, and computer-communication services, a dramatic change from the previous separate industry structures and separate policy treatments. In Canada convergence has also established a framework for competition between the telecommunications carriers and the cable-televison companies. The CRTC introduced telecom and broadcasting policy changes in 1998 that permit the telephone carriers to offer cable service; conversely, the cable companies may now compete by providing local telephone services. In addition, the Broadcasting Act was amended to remove the prohibition on Bell Canada and the other telcos from holding a broadcasting distribution licence.

As Babe argues, the move towards convergence policy has resulted in more mergers and amalgamations across what previously were distinct industry boundaries. He goes even further and charges the then head of the CRTC, Commissioner Spicer, with having blind faith in the benefits of technological outcomes (1995: 294–5). Winseck, however, notes that convergence has less to do with technical feasibility and more to do with politics, power, and broader cultural policy objectives (1998: 188). In addition there has been political support from business, new communications and media players, governments, and a group known as the Coalition for Public Information (the CPI is no longer in existence) for convergence (Moll & Shade, 2001).

In August of 1996 the federal government issued its convergence policy statement, establishing broad policy objectives for telecommunications and broadcasting in the context of the information highway. Essentially, convergence policy removes the prohibition on Bell Canada from holding a broadcasting licence and permits competition between the telecom carriers and the cable-TV companies in their core markets (Industry Canada, 1998: 49).The impact of these developments on the information highway and the government's desire not to regulate the Internet or new media services is not surprising given NAFTA, which places limits on or makes any broad cultural policy obligation virtually impossible (CRTC, 1999c; Winseck, 1998: 313). Even more significant is international multimedia policy, as articulated in the World Trade Organization (WTO). The WTO goes further than the service sections of CUSFTA and NAFTA by targeting communications, culture, and computer services. Communications services range from audio-visual to postal to telecommunications; cultural services cover entertainment, news agencies, libraries, and museums; computer services entail consulting, computer-hardware installation, software implementation, and data processing (WTO, 1998b). The more immediate concern of the WTO is the heavy regulation of audio-visual services by most countries. The WTO has recommended that competition, liberalization, and deregulation be extended to audio-visual and multimedia services in order to move away from the cultural exemptions in trade agreements or other cultural-protection policies (WTO, 1998a: 6). This international neo-liberal policy approach, when combined with global electronic commerce, is aimed at opening up markets further to ensure the free flow of data, information, and audio-visual content over the Internet with as little state interference as possible.

FEDERAL GOVERNMENT REORGANIZATION

A neo-liberal policy approach also requires that the size of the state be reduced, and that its role in the economy be greatly diminished (Schultz, 1988; Globerman, Stanbury and Wilson, 1995). One of the large telecommunications-user organizations, ITAC, made a strong recommendation in the early 1990s that the government of Canada act on its recommendation to "move swiftly to streamline federal government operations by reducing the number of government departments by one half" (ITAC, 1993: 3). ITAC and other telecommunications-user organizations were part of the procession of pro-business lobby organizations, such as the Business Council on National Issues, the Canadian Chamber of Commerce, the C.D. Howe Institute, and the Fraser Institute, which argued that the federal government needed to be downsized. By the mid-1990s the Treasury Board Secretariat was structurally reorganizing the state as indicated in the Federal Regulatory Plan 1994 (1993). Under the plan the Department of Communications was eliminated and a new department created called Canadian Heritage, which included components of the former DOC covering the Broadcasting Act, the Department of Communications Act, and the CRTC Act, as well as other acts that dealt with culture and the cultural industries (Treasury Board, 1993: 25). Telecommunications, however, was "taken over" by the enlarged department Industry Canada,[2] which is now responsible for a broad range of acts that include the Board of Trade Act, the Canada Business Corporations Act, the Canada Corporations Act, the Competition Act, the Competition Tribunal Act, and the Telecommunications Act among others (Treasury Board, 1993: 147). The mandate and guiding principles of Industry Canada that affect telecommunications policy are now based on the "fair and efficient operation of the Canadian marketplace" (ibid.: 146). Another important objective that applies to telecommunications is to provide industries with the legislative and regulatory infrastructure to compete in the global marketplace.

FREE-TRADE SUPPORT AND RESISTANCE

In response to the Canadian economic recession of the early 1980s, opposition to the perceived interventionist federal liberal policies into the economy, and growing concern about United States protectionism, business lobbies led the way as political supporters of continental free trade. Political activists and supporters of free trade between Canada and the United States included the Business Council on National Issues (BCNI), the Canadian Chamber of Commerce, the Canadian Federation of Independent Business, and the Canadian Manufacturers' Association.[3] The BCNI and Chamber of Commerce, in particular, with their counterparts in the United States, the American Business Round Table and the U.S. Chamber of Commerce, lobbied their governments for a bilateral trade agreement and the establishment of a North American trading bloc as a way to gain more leverage against the Asian and European blocs. The proposed trade agreement would remove continental barriers to investment and intellectual property, and would add, for the first time, liberalization of trade in services. These political pro-free-trade agents and others such as the American Coalition for Trade Expansion, the C.D. Howe Institute, the Fraser Institute, the Economic Council of Canada, and a spin-off lobby of the BCNI, the Canadian Alliance for Trade and Job Opportunities,[4] advocated throughout the 1980s and 1990s the benefits that would accrue to Canadians once a free-trade agreement was reached with the United States. As a point of interest, free trade was also supported by the Consumers' Association of Canada (CAC). A former employee of the CAC, David McKendry, stated that the organization supported unrestricted imports and free trade so that consumers could have access to products at cheaper prices (interview, D. McKendry, June 1997). This continentalist position, based solely on individualism and consumption, ignored the negative aspects that continental free trade would have on social policies, as well as the loss of many telecommunications jobs.

Most Canadians, however, were not supporters of the free-trade initiative. Anti-free-trade forces (*The Other Macdonald Report*) mobilized in 1985, forming two organizations, the Council of Canadians and the Pro Canada Network, renamed the Action Canada Network. The Action Canada Network (see Appendix 20 for a list of members) is a broad coalition that includes experts, academics, women's organizations, religious organizations, labour, seniors, students, public-interest groups, professionals, and native peoples' organizations from all parts of the country. Both organizations informed Canada's citizens that free trade would result in a loss of Canadian sovereignty and that social policies that dealt with health, education, welfare, unemployment, job training, regional policies, and the environment would be challenged. Concerns about severe job losses and the loss of democracy were also raised in various publications by the Centre for Policy Alternatives and the Council of Canadians.

The Council of Canadians was involved in educational awareness and coalition-building among local grassroots organizations promoting citizen involvement to counter private forces and the neo-liberal remodelling of Canadian national policy as well as its social system.[5] In the council's view, free trade would deepen class disparities (Council of Canadians, 1993b: 3; interview with Dave Robinson, then director of research, 1995). Both organizations were successful in terms of informing the Canadian population that the major beneficiary of free trade would be big business.

The governing Progressive Conservative party called a federal election in the fall of 1988. Often referred to as the "Free Trade Election" it was the culmination of more than a decade of debates over development strategies and trade issues. The election took on the form of a "national referendum on free trade, suggesting that Canadians would cast their ballots on the basis of a single issue rather than the more familiar appeals of leadership, ethnicity and regionalism" (Brodie, 1989: 175). As Brodie explains, during the election the capitalist class resorted to expensive efforts, conducting more than twenty-five national and private polls to track the party's progress on the free-trade issue (181). Despite the millions of dollars spent by big business, approximately 58 per cent of the electorate voted against the incumbent Conservative government and its support of the free-trade mandate. With only 42 per cent of the popular vote, and the Canadian federal parliamentary system of first-past-the-post, the Conservatives were voted back into office. Their platform of trade liberalization was initiated soon after by two successive agreements, the Canada – u.s. Free Trade Agreement (CUSFTA, 1988) and the North American Free Trade Agreement (NAFTA, 1992). Both initiatives extended continental liberalization of investment and trade in services as well as goods.

Throughout this period the Department of External Affairs, now the Department of Foreign Affairs and International Trade, was the instrumental state agent and organizing vehicle to create hegemonic consent for CUSFTA, NAFTA, and the World Trade Organization. As early as 1986, two years before the free-trade election, the minister of External Affairs announced that an International Trade Advisory Committee would be headed up by Walter Light, then CEO of Northern Telecom; Wendy Dobson, the deputy minister of Finance and former head of the C.D. Howe Institute; and Richard Lipsey, an academic economist who also played a key role in the pro-free-trade lobby (Doern and Tomlin, 1991: 110). Initially set up as a neo-corporatist committee that would operate based on consensus, the committee was to "ensure that we pursue domestic policy initiatives that are complementary rather than antagonistic to trade policy objectives. Canadians – governments, business, labour and the academic community – need to cooperate and work toward a common agenda in order to forge the new relationships required to meet the competitive challenges of the global economy. The

challenge is to face future negotiations with a greater degree of pre-
paredness ... from the GATT, FTA, and NAFTA negotiations" (ITAC
1993a).

The International Trade Advisory Committee set up sixteen Sectoral
Advisory Groups on International Trade (SAGITs) to complement the
work of the committee. These groups were very important in providing
sectoral viewpoints on all matters in the CUSFTA and NAFTA negotiations. Sectoral groups included agriculture, fishing, mining, energy,
chemicals, forest products, automobiles and aerospace, general and
financial services, and information industries, as well as a SAGIT on
information technologies and telecommunications (see Appendix 21
for a complete list of trade committees and sectoral advisory groups).
Jim Kerr, the International Trade SAGIT director for information technologies and telecommunications, pointed out that most of the representatives were from the private sector (interview with J. Kerr, 1997;
ITAC and SAGITs background; Department of External Affairs, Communiqué, 1986). A closer examination (see Appendix 22) of current
and past members reveals that these representatives were also chief
executive officers and presidents and vice-presidents of private corporations, with only token communications-labour members. Members
were selected by the Department of International Trade. Once selected,
members sat on the SAGIT for a two-year term, which could be renewed
for successive terms.

Kerr described the ITT SAGIT as a high-level advisory committee that
interacted directly with the minister, engaging in a two-way flow of
information with the federal government. It allowed the private sector
to be privy to the free-trade negotiations as well as to provide input.
Membership also required the signing of a security declaration to make
sure that information exchanges remained confidential and classified.
The exclusionary and secretive aspects of the advisory committee were
obviously designed to prevent public input or accountability to the rest
of the Canadian population. Rod Hiebert of the TWU, a member of
the ITT SAGIT, noted that it was preoccupied with how to get other
countries to open up their markets for business to sell technology and
technical expertise. Concerns about labour issues and job losses were,
however, consistently ignored. Hiebert disclosed that most of the SAGIT
members believed that free trade in services would lead to more job
loss in the telecommunications sector (interview with Rod Hiebert,
February 1996; comments by TWU President R. Hiebert at the Toronto
SAGIT, May 1996). Needless to say, current SAGIT meetings are preoccupied with the fall-out from these neo-liberal trade policies and their
effect on the work force, employment, education, and training (J. Kerr,
June 1997).

Both trade agreements have helped to advance convergence, the information highway, and the Internet. As noted by Babe, information and knowledge have historically been treated as part of public goods and community resources (1995). The information highway, the Internet, and the World Wide Web raise social and civil issues because of the pay-per aspect imbedded in the services that these webs and networks offer (Mosco, 1989). As continental neo-liberal policy extensions, CUSFTA and NAFTA have not only extended economic activity to more aspects of social and cultural life; they also fortify a market form of policy regulation, which, in the case of enhanced, value-added, and multimedia services as well as the Internet, is a zone that is free from public accountability or oversight.

FREE TRADE AS TELECOMMUNICATIONS POLICY

Both the Canada u.s. Free Trade Agreement and the North American Free Trade Agreement fulfil part of the wider neo-liberal agenda. Both trade agreements also advance the neo-liberal hegemonic project, contributing to a policy paradigm shift. Ostensibly, these policy changes in telecommunications created competitive markets. As this chapter and previous chapters reveal, as a capitalist state the Canadian federal government was very active in creating a market economy in this sector, allowing new telecommunications entrants into the market and relaxing, or requiring only light regulation over the telecommunications service sector. At a broad policy level the federal government also redefined social relations as market relations, emphasizing their benefits for users and consumers.

The major objectives of both CUSFTA and NAFTA included reducing tariff and non-tariff barriers, strengthened corporate rights, and relaxed restrictions on corporations and multinationals regarding continental activity in mergers, takeovers, investment, and capital mobility. Viewed from this vantage-point, the move towards freer trade "represents both continuity and change" (Breton and Jenson, 1991: 206). Certainly both trade agreements are a continuation of the institutionalization of postwar strategies to deepen and extend "market forces." In addition, both trade agreements are a continuation of past and ongoing struggles. Concurrently, CUSFTA and NAFTA mark a break for Canada, a move away from the notion that the federal government should promote a national development strategy. As Brenton and Jenson write, CUSFTA "marked Canada's allegiance to neoliberalism and enthusiasm for 'market-forces' as well as its integration into a North American bloc" (1991: 205–6). While CUSFTA and NAFTA are a continuation of past social practices, they are also new because they now locate Canada within a continental institutional framework. This institutional framework situates the Canadian, American, and Mexican experience within a global context as well as a continental one. It has resulted in additional policy layers that must be taken into consideration, layers that can supersede national policies. Specific elements, such as Canadian telecommunications policy changes, are the result of changes brought about by the social relations of domestic and foreign multinationals in their interaction with the federal government in order to respond to a new reality of expanded worldwide trade. Part of this adjustment involved a shift to institutionalized continental policy arrangements.

Article 102 of CUSFTA puts forward the objectives and scope of the first broad continental agreement. Essentially these objectives strengthen economic rights for multinational businesses by eliminating barriers to trade in goods and, for the first time, adding trade in services as well; by facilitating fair competition; through liberalization within the trade area; through the establishment of efficient joint administrative procedures; and by laying the foundation for future bilateral and multilateral arrangements (Canada, 1988: 9). Articles 103 and 104 extend these obligations, in Canada's case, to national, provincial, and local governments by giving CUSFTA precedence over these other levels of government. Articles 105, 501, and 502 entrench national treatment with respect to investments and trade in goods and services (10, 197). For telecommunications, this means that all service providers, whether wired or wireless carriers, domestic or foreign resellers, are to be treated equally – that is, national treatment must be offered to telecommunications businesses even if their link is only electronic.

The ground-breaking aspect of the CUSFTA is the introduction of trade in services. Trade in services is considered by capital the new frontier of international commercial policy. Chapter Fourteen of CUSFTA establishes the rules of trade in services. This chapter addresses the point that services are increasingly mingled with production, sales, and distribution as well as the service of goods. Moreover, reliance on advanced communications systems is required to facilitate trade in services. Consequently, Chapter Fourteen removes the barriers in both countries to trade in services. This chapter was developed simultaneously with the General Agreement on Tariffs and Trade, which recently introduced the General Agreement on Trade and Services (1994) (interview with B. Tritt, Stentor, formally the chief communication negotiator for the Department of Communications on CUSFTA, June 1997).

Chapter Fourteen deals with services, investment and temporary entry. The services are listed in Annex 1408 of the chapter including telecommunication networks, enhanced services supporting open and competitive markets. Article 3 establishes just how these markets are to operate in terms of:

a) access to, and use of, basic telecommunications transport services, including but not limited to, the lease of local and long distance telephone service, full-period, flat rate private line services, dedicated local and intercity voice channels, public data network services, and dedicated local and intercity digital and analog data services for the movement of information, including intra corporate communications;

b) the resale and shared use of such basic telecommunications transport services;

c) the purchase and lease of customer premise equipment or terminal equipment and the attachment of such equipment to basic telecommunications transport networks;

d) regulatory definitions of, or classifications as between basic telecommunications transport services and enhanced services or computer services;

e) subject to Chapter Six (Technical Standards) standards, certification, testing or approval procedures;

f) the movement of information across borders and access to the data base or related information stored, processed, or other wise held within the territory of a Party. (211–12)

Sections 2 and 3 of Article 3 explain what is meant by a commercial presence and investment. Article 4 ensures that these commercial agents have access within and across borders for enhanced services. Article 5 ensures that basic telecommunications transport facilities can continue even for a monopoly, but with strict rules that establish how a monopoly can act – that is, like a competitive enterprise. The telecom monopoly cannot engage in any anti-competitive behaviour, including predatory activity or cross-subsidization, or deny access to its basic telecommunications networks or services (211–12).

The CUSFTA has put more pressure on Canadian domestic policy to create a competitive telecommunications market. The 1988 Free Trade Agreement adds another policy layer that takes precedence over Canadian domestic policies to ensure that the conglomerates and transnationals operating in the North American bloc have unrestricted access to basic telecommunications networks and telecom business services. The main objective of the transnational and conglomerates was to ensure that there would be no impediment to the free flow of information, voice or data, between the United States and Canada.

Although Article 2010 of CUSFTA states that "nothing in this Agreement shall prevent a Party from maintaining or designating a monopoly," the agreement seems to have influenced the Canadian federal and provincial governments to rethink their positions concerning Crown corporations or public monopolies (299–300). Responses from both levels of government have included the privatization of the federal Crown corporations Teleglobe and Telesat. The provincial governments of Alberta and Manitoba privatized their Crown corporations, Alberta Government Telephone (now Telus) and Manitoba Telephone System respectively. Article 1605 prevents either level of government from ever establishing a public monopoly again. Once privatization has taken place, for the re-establishment of a Crown corporation there has to be "payment of prompt, adequate and effective compensation at fair market value" (236). These severe measures have certainly placed federal and provincial Canadian governments on an integrative policy path of telecommunications continentalization with the United States.

The North American Free Trade Agreement (NAFTA) broadens telecommunications continentalization further to include investments, intellectual-property licensing, standards, and what the document calls transparency of rules and regulations. As CUSFTA before it, NAFTA identifies and recognizes the centrality of telecommunications services in the contemporary global economy. What has become critical to corporate needs is unhampered access to public telecom networks, to transfer voice and electronic data across borders in North America. Consequently, NAFTA devotes its Chapter Thirteen to establishing rules for cross-border telecommunications service and investment, in the process expanding the telecommunications commitments in CUSFTA. Article 1302 of NAFTA establishes the protocol that allows business enterprises to have access to and the use of any telecommunications network or service within a country or across the borders of Canada, Mexico, and the United States to conduct business (Canada, 1992: 13–4, 230). Article 1302, sections 2 (a) through (c), identify the rights that business enterprises now have, including the right to purchase, lease, and attachment of their terminal equipment or other equipment with the public telecom networks in each country; to interconnect their private leased or owned circuits with the public telecommunications networks within each country or across borders; and to perform their own switching, signalling, and processing functions. In essence, the most important aspect of Chapter Thirteen affecting telecommunication policy is the right it gives to companies to access public networks in order to move information freely across North American borders through their own privately secured data bases.

Moreover, NAFTA also ensures that the rates the conglomerates and transnationals pay to access these public telecommunications service networks will be based on economic costs; that is, the costs directly related to providing a telecommunications service such as basic local service are free from cross-subsidies (Canada, 1992: 13–1, 227; Janisch, 1989: 99; Schultz, 1995: 280). Where CUSFTA was ambiguous about the economic costing method to be used to access public networks, NAFTA establishes a method similar to the one used in the United States of cost-base pricing, which claims to eliminate all subsidies from long-distance telephone service to basic local service and from residential urban basic service to rural. Essentially, this economic-costing provision of Chapter Thirteen permits the institutionalization of reverse cross-subsidization to the large telecommunications users, whether they are multinational corporations or government users, at the expense of all the citizens in all three countries.

Another important aspect of NAFTA is the investment chapter. Chapter Eleven on investment builds on CUSFTA by imposing more extensive obligations and by creating a process for the binding settlement of investment disputes between NAFTA investors and the participating NAFTA governments. Fundamentally, the NAFTA investment chapter establishes common rules for the treatment of investment from investors of other NAFTA countries; it liberalizes existing investment restrictions; and it provides a mechanism to resolve investment disputes between investors and other NAFTA governments (Canada, 1992: 11–1, Article 1101). NAFTA does not, however, end government requirements that a certain percentage of corporate directors be nationals. What it does provide are limits on government regulations and that the senior management of any state enterprises can be Canadian, Mexican, or American.

It would appear that liberalizing investment regulation would allow telecommunications companies from Canada, Mexico, and the United States to invest more freely in each other's countries. But an interview with Bell Canada's vice-president of law and regulatory matters, Bernard Courtois, reveals that policy and legislative changes that include the long-distance telecommunications-competition decision (CRTC 92–12), the Telecommunications Act, CUSFTA, NAFTA, and now the World Trade Organization liberalization regarding telecommunications investment of 46.7 per cent make Canada a very open market to invest in. By contrast, although telecommunications investment in the United States appears to be open, a company has to get permission first from the Federal Communication Commission, then permission from each state. Not only is the process very costly; it also takes a number of years (interview with B. Courtois, June 1997). American telecommunications companies have dramatically increased investment since liberalization. What Bell Canada and the Stentor member companies objected to are policy, trade, and legislative changes that treat foreign competitors favourably by not placing any regulatory restrictions on the foreign companies. In Stentor's view, its members were hampered by regulatory constraints and favourable investment opportunities in the United States (interview with Bob Tritt, national director of international affairs, Stentor, 1997).

Because NAFTA was negotiated by the chief executive officers of major transnational corporations from Canada, the United States, and Mexico, as well as the government representatives of the respective countries (Japan Economic Institute, 1992, No. 39A), it has been touted as "a vehicle to do business more efficiently" (Shefrin, 1993: 14). It has prompted critics such as Chomsky to call NAFTA "an executive agreement" that was secretly negotiated, given that the governments of each country refused to make drafts of the agreements available for public comment (1993: 414). This format of neo-liberal policy-making helped to silence opposition and contain resistance by excluding the oppositional forces. NAFTA and CUSFTA as well as the World Trade Organization create new continental and international government institutions that add additional telecommunications policy and regulatory layers. All the practices of these institutions are undemocratic and exclusionary, shifting decision-making and regulation of telecommunication policy to an unelected elite group of business and government agents, whereas the people most affected by these agreements, citizens such as telecommunications workers, consumers, and public-interest groups are not even allowed to intrude in this narrowed "public arena," particularly at the continental and international levels. The opposing forces will have to carry their resistance into the continental and international arenas and forge alliances with other groups, non-government organizations, and unions globally.

CONTINENTALIZATION AND LABOUR

The Canadian telecommunications unions responded to neo-liberal policy continentalization by reviving the Canadian Federation of Communication Workers. At a national Canadian conference of communications unions in 1992, an informal alliance was forged among the major unions (see Appendix 23 for a list of members). The Alliance, as it is called, was established as a response to the Stentor reorganization and the devastating effect that was expected in terms of union job losses resulting from the liberalization of foreign-investment rules. In the Alliance's statement of unity, a common program was established to co-ordinate information from the various telecom industries, present a common union position to the Stentor member companies, and establish a common position for contract negotiations (interview with S. Shniad of the TWU, 1996, and C. Simpson, ACTWU, 1996). Rather than being a separate, formalized structure, the Alliance is a political agent, one that has been active in fighting against NAFTA. Other unions such as the TWU have forged wider alliances and coalitions with labour and other social organizations domestically, continentally, and internationally.

Continental alliances have also been formed by the largest telecommunications unions in Canada, Mexico, and the United States (see Appendix 24). In December 1991 union presidents Fred Pomeroy of the Communications, Energy and Paperworkers Union of Canada (CEP), Francisco Hernandez Juarez of Sindicato de Telefonistos de la Republica Mexicana (STRM), and Morton Bahr of the Communications Workers of America (CWA) signed on to the first permanent continental labour coalition. Their agreement focuses on strengthening the ability of each union to bargain, support joint mobilizations if there is a strike, and defend unions' and workers' rights in alliance with the international trade secretariats and other social movements in each country.

This alliance is all the more extraordinary because the three unions did not agree on NAFTA. The CWA and CEP, along with the Canadian Alliance members, strongly resisted the free-trade agreement because it would contribute to the loss of unionized telecommunications jobs. The Mexican union leadership, however, supported NAFTA. STRM leadership, like the Mexican government and private sector, supported NAFTA because all three agents believed that the investment chapter of NAFTA would permit foreign investors to inject money to modernize Mexican telecommunications. STRM saw this as a way of creating more union jobs. This is why STRM also supported the government's sale of TELMEX, the publicly owned and operated telephone company, as well as the technological upgrading required to make the company more marketable (K. Moody and M. McGinn, 1992: 69).

Since the signing of NAFTA there has been job growth in the Mexican telecommunications industry, but not at TELMEX. Some growth has occurred in the non-unionized areas of wireless communications and telecommunications services. Major job losses of 27,000 TELMEX employees have occurred but have been explained away by TELMEX as a result of technological upgrading and re-engineering[6] (STRM, TIE-North America, 1994: 8). Moreover, these changes have produced an opposition movement within STRM. In 1991 about fifty activists from the democratic movement within STRM formed a smaller union in Baja, California. This breakaway union is now a member of the Postal Telegraph and Telephone International (PTTI); appendix 25 lists the North American and Caribbean affiliates. PTTI activities include extensive published research on conditions of employment and job-related problems of telecommunications industries, research that places an emphasis on transnational telecommunications activity.

At the Transnationals Information Exchange (TIE) North American conference in 1994, participating unions such as the CWA, CEP, TWU, STRM, and Baja California and Sonora Telephone Workers Union shared information concerning the result of the implementation of NAFTA. Recent developments include the rapid introduction of new technology and major industrial restructuring and its impact on workers in all three countries. The TIE conference found that there are similarities in plant closings, industry downsizing, and the slashing of wages and benefits, and that work continues to be deskilled. Conference participants agreed to open new channels of communication, widen the focus of trade-union activity beyond domestic concerns, and develop strategies that will challenge the power of the multinational telecommunications corporations (TIE-North America, 1994: 1).

With the exception of a CEP strike in the early 1980s and a strike by the Communication Workers of America at Bell Atlantic in 1989, the telecommunications industry benefited from fifteen years of labour peace, commencing in the mid-1970s. Competition and extensive reorganization of the industry through mergers has resulted in concentration of telecommunications ownership and the revival of tele-communications-union militancy. This new tide of North American militancy and mobilization began in Puerto Rico with a forty-day strike in June 1998, followed by a two-day general strike in opposition to the sale of the state-owned Puerto Rico Telephone Company to GTE.

In the United States, after the introduction of the 1996 Telecommunications Act, the industry engaged in merger mania with a vengeance, which resulted in the newly formed SBE, formerly Southwestern Bell and Pacific Telesis; Bell Atlantic, formed from the Bell Atlantic, Nynex and GTE merger; the AT&T merger with TCI; and the World Com.–MCI merger, among others. Once the mergers were completed, these newly formed companies engaged in major downsizing, particularly of union jobs. As Kim Moody points out, depending on the company, after downsizing, 30 to 60 per cent of union jobs were eliminated. The overall union rate for the industry fell from 55 per cent in 1983, prior to the AT&T break-up, to 27 per cent in 1997 (Moody, 1998). By 1998 the industry workers' peace ended as the Communication Workers of America and the International Brotherhood of Electrical Workers struck various locals over issues such as forced overtime, increases in temporary jobs, contracting out, pay-for-performance (work flexibly), required overtime of twelve to fourteen hours at a pay rate of straight time, and low staffing levels attributable to new digital services. Responses involved new strike tactics such as flying squads or mobile pickets, as well as an active campaign to get AFL-CIO members and telephone subscribers to exercise their rights as consumers and switch carriers.

Downsizing has been so severe at U.S. West in Portland, Oregon, that not enough personnel were on hand to deal with quality-of-service problems. Telecommunications continentalization is also evident in labour-force relations, as TWU technicians went to Oregon to help upgrade the networks and make repairs (S. Shniad, 1999).

Labour militancy has also been on the rise at the Canadian telecommunications unions. Bell Canada asked for concessionary demands from its 2,300 telephone operators. Bell wanted to relocate 1,400 of the operators who provide directory and relay assistance and reduce their wages from $19 an hour to $10. Relocation would be to an operator-service company created by Bell in partnership with Excell Global Services of Arizona. This continental contracting-out would leave Bell with 900 operators in Ontario and Quebec. As Gary Cwitco, the national representative for CEP, stated, the move to contract out operators to other North American locations was not because of unprofitability. Throughout 1997–98, operator services contributed 12 per cent to Bell's profits.[7] Lowering operator wages, eliminating benefits, and establishing temporary operators permits the new operator-service company to contract these services to the new telecom service providers.

Bell's savings are also unlikely to be passed on to their customers in lower service rates. People in Bell's operating territory have faxed, phoned, and written complaints to the CRTC and Industry Canada about how the company is treating its low-paid female workers. Although Bell and its shareholders have downsized and contracted out, the company seems to have forgotten the importance of front-line contact with subscribers and the company goodwill that is developed through "real" operators and technicians. Poor or negative customer service in a competitive environment often results in punishment, as the cable industry found out over the issue of negative billing.[8] A number of irate customers gave up their service or switched to satellite dishes.

Bell executives gambled that the ensuing operators and technicians' strike could be easily handled by company managers, who signed up for sixty-hour-a-week strike duty and pay premiums. While communications analysts such as Ian Angus agree that on the surface Bell was able to take care of operator requests and repairs, he cautions that competition means customers can be less forgiving if Bell can't deliver their service requests and gives them a choice to take their business elsewhere.

8 Conclusion

This book provides a comprehensive examination of the major Canadian telecommunications policy changes that have occurred over the last twenty years. It examines the primary capital and state agents who were instrumental in initiating the policy shift to a neo-liberal telecommunications model, and the centralization of telecommunications power with the federal government. I have foregrounded the tensions and conflicts that resulted from the establishment of neo-liberal telecom policy changes. For more than two decades, opposition and resistance did not let up. From the beginning this policy shift was met with strong opposition at many levels, but its strongest resistance came from labour, consumer, and public-interest organizations. These opposing forces have been strengthened through class and social alliances and coalitions. Because of significant internal disagreements, however, these alliances were not successful in presenting a united front to prevent long-distance and local telephone competition; to establish a lifeline program for low-income Canadians; or to participate in the free-trade negotiations. Consequently, federal government departments and the CRTC were able to take advantage of the divisions within the opposing forces.

Although this neo-liberal telecom policy shift is a continuation of past liberal practices, it also advances new ones. Many of the policy changes continue the post-war strategy of extending market forces by deepening and extending capitalism continentally and internationally. The federal government did this by retreating from modern liberal policies, returning to a previous form of liberalism, one with fewer controls, without tariff restrictions, and little or no regulation of capital activity. The continental regime has been aided by the federal government and multinational corporations to ensure the integration of telecommunications policy. The primary purpose of the continental initiative was to change existing policy ideas and regulatory practices so that capital could benefit from expanding continental and international markets.

The new, hence "neo," aspect of this liberalism adds another policy layer, a continental layer – through bilateral and trilateral free-trade agreements – to the existing national and provincial ones. These continental agreements are significant because the policies they generate supersede any national, provincial, or local policies, making these agreements very powerful. The continental trade agreements, the Canada-u.s. Free Trade Agreement (CUSFTA) and the North American Free Trade Agreement (NAFTA), provide the substance for part of this policy layering. Another new part of this layering is trade in telecommunications services, specifically enhanced and value-added services. The second agreement, NAFTA, removes more barriers to capital investment and introduces new rules and regulations for intellectual-property rights. Recognizing the increased importance of telecommunications services to continental economic development, NAFTA devotes an entire chapter to establishing new policy governing this sector. Both trade agreements locate Canada further within a continental and international policy framework.

New legislation such as the Telecommunications Act contributes to this continental policy layering in that it works in conjunction with the trade agreements. This domestic telecommunications legislation furthers continental processes and practices by foregrounding economic rights and emphasizing that Canada must rely on market forces. It also paves the way for further continental corporate integration in telecommunications by increasing foreign-ownership rules. Other benefits include the concentration of telecommunications power and control with the federal government by transferring jurisdiction and regulation of all telecommunications away from municipal and provincial governments to the federal government.

The development of a neo-liberal, continental telecommunications model has benefited large corporate users, the new competitors, and the established dominant providers. Corporate rights are strengthened within the North American continent through CUSFTA and NAFTA. Both agreements ensure that multinational corporations will have access to and use of any telecommunications networks or services so that they can exchange business information freely, be it voice or data, across the borders of the members' countries. Multinational corporations headquartered in Canada, Mexico, and the United States have been guaranteed that the telecommunications market will prevail and that the rates these corporations pay will be based on economic costs, devoid of most subsidies that may try to address social needs. Telecommunications multinationals operating in these countries can further integrate their businesses because all three countries have weakened their previous foreign-investment restrictions.

Telecommunications competitors have benefited by entering the market, obtaining market share, and they can exit the market whenever they wish, often leaving consumers stranded. By and large the new competitors benefit by not having their activities regulated by the state. The established telecommunications service providers have also benefited from neo-liberal policy changes: despite their loss of some market share, with restructuring their profits have continued to rise. What is also striking for the dominant carriers is that they are now subject to very little state regulation. In addition, they have been released from concern for much of the "public-good" aspect of a telecommunications utility, primarily universality and affordability objectives that ensured that all telecommunications subscribers were treated fairly and equally.

Yet the cost of the neo-liberal telecommunications policy shift has been born by the majority of Canadians. One of the outcomes of following this policy agenda has been massive employment reductions in the telecommunications sector. A large number of telecommunications workers in Canada, the United States, and more recently Mexico have lost their jobs. Most of these are union jobs that, until recently, paid a decent wage and benefits. New jobs created in this sector have been largely part-time or contracted-out. Although the new telecommunications competitors have created some employment, most of the work is non-union and low-paying. Furthermore, the jobs created by the competitors do not offset the jobs cut by the established telecommunications carriers. In a neo-liberal environment, where all telecommunications competitors fight for market share, profits, and return on investment for their shareholders, corporate objectives involve keeping all expenses low, whatever the cost. Labour expenses then become an easy target.

The neo-liberal telecom policy changes also redefine the "public interest" to mean more choice for the large telecom consumers, who can now choose from different service providers and products. This has had, however, a devastating effect on many residential telecommunications subscribers, particularly those with low or fixed incomes and those who live in the very remote regions of the country. Only economic criteria were taken into consideration regarding these consumers/users, and their importance is measured by individual consumption capabilities. These standards create a hierarchy or stratification of consumers. At the top of the hierarchy are the large business users/ consumers, who are best able to take advantage of long-distance service discounts, bulk attachment rates, and the array of new technological products and services. Next come the upper-class and middle-class consumers. With more disposable income, they are able to take advantage of long-distance discounts and to subscribe to and pay for the vast array of new services. At the bottom are those who are least able to afford competition, choice, and new services: the poor, the working poor, the unemployed, those on fixed incomes, such as seniors and the disabled, indigenous peoples, and others living in the rural or remote areas of Canada. These are the people who, as a result of neo-liberal telecommunications policy changes, now pay on average 81 per cent more for their basic local service than they did in 1992.

Concurrently, the previous government affordability objective, based on a broad system of cross-subsidization, has been narrowed or targeted to those most in need. Although many countries have introduced some form of telecommunications targeted-subsidy program for low-income subscribers, the Canadian neo-liberal telecom policy changes to date have not. The Canadian regulator is on record as stating that affordability is not a problem and such a subsidy is not necessary. What has been substituted instead is an affordability-tracking program. Ironically, affordability is being tracked by the Stentor member companies, Bell Canada, and the provincial telephone companies. This research tells us that there are a significant number of people, approximately 250,000, who have had to give up their basic telephone service because they can't stretch their incomes any further. At the time of this writing the CRTC has indicated to consumer and public-interest advocates that it would like to address the affordability issue for very low-income Canadians. The affordability issue has major implications as the country embraces policies moving it towards an information society/knowledge economy, wherin access to the Internet and the World Wide Web, through wire or wireless systems, is increasingly becoming very important.

Neo-liberal telecommunications policy is characterized by the rule of the market and policy changes that include liberalization, neo-regulation, privatization, and continentalization. Policy liberalization helps to create markets where before only telecommunications monopolies existed. The federal government changed existing policies to allow new competitors to enter the long-distance and local service markets to compete with the established telecommunications providers. But eliminating the monopolies in the long-distance and local service areas has not really resulted in a competitive marketplace. None the less, the established telecom corporations such as Bell

Canada and the provincial telephone companies, for example, continue to dominate the market. Although other, smaller telecom companies have obtained some market share, this has not resulted in a competitive market. Instead, the monopoly environment has been replaced with a telecommunications oligopoly.

Neo-regulation rather than deregulation better explains the nuances of regulation in the new telecom policy environment. Where dominant telecommunications providers prevail, they continue to be regulated by the state, but only lightly. In other areas, where a competitive market has been created, such as business services and telecommunications equipment, state regulation has been replaced by market regulation or self-regulation. New Canadian telecommunications legislation also allows the communications regulator to forbear from regulating if it deems a market competitive enough. Although forbearance does not explicitly describe what happens if market failure were to occur, it implies that the state may step in to re-regulate.

Privatization policies have, with one exception in the province of Saskatchewan, ended Canada's mixed public and privately owned telecommunications system. Federal Crown corporate holdings for the international telecommunications network and satellites have been sold to the private sector. The two provincial governments of Alberta and Manitoba have also privatized their telecommunications companies. Other Crown corporations such as Teleglobe and Telesat are now private corporations. In addition, Industry Canada has begun to auction spectrum to the private sector.

While liberalization, neo-regulation, and privatization reorganize domestic policies, continentalization introduces another policy level through bilateral and trilateral trade agreements to address the continental telecom's external activity. Although these agreements are defined as free-trade agreements and supposedly make the rules and regulations for continental trade more transparent, they are primarily limited to economic issues. Both CUSFTA and NAFTA broaden the rules governing the trade of goods and, for the first time, introduce trade in services. They also temper restrictions on foreign investment, establish the framework for intellectual-property rights, and apply telecommunications standards and definitions established in the United States to Canada and Mexico as well.

Telecommunications policies established through these continental agreements supersede national or provincial ones. This means that domestic legislation and regulation must reflect the continental neo-liberal thrust. Consequently, economic rights and objectives have been foregrounded to address the market needs of multinational corporations so that goods, services, and information can move freely across the borders of the members' countries. At the same time, continental-ization downplays and ignores social or cultural objectives and rights.

Continentalization places the rights of multinational corporations above the rights of citizens and states. Continentalization as a social practice is at once inclusive and exclusive. Both continental agreements were made in secret by state and business policy agents. This procedure provided a way to silence the extensive opposition there was in Canada to neo-liberal policy changes, an opposition that was hampered by the absence of any state institutions or structures to accommodate political participation from interested social agents. These continental agree-ments excluded all other voices and points of view whose concerns were social or cultural rather than economic, such as those of labour, public-interest advocates, consumer organizations, and community groups. These exclusionary factors reveal the incompleteness of the project, and a possible area of regime vulnerability.

This research has concentrated in a unique way on the important role the federal state played in implementing a neo-liberal telecommu-nications policy model, and on the incompleteness of the regime. It reveals that the federal government was not a neutral player and that its primary concern was not just the management of conflicts among competing capital elites. It also shows that the state did not implement neo-liberal policies because it was an agent for the international-capital class. The research shows that the state was an important actor, along with business and their political lobby organizations, in producing hegemonic consent for the continental, neo-liberal telecommunications policy model. However, this consent is based on a limited hegemony, the result of not garnering support from the entire population, or granting very few material concessions. Rather, the neo-liberal project has contributed to and intensified resistance and opposition to it. There are two major reasons why the project has increased resistance. First, because it is primarily concerned with service sectors such as transpor-tation, finance, and telecommunications, the cost of the project has been passed on to the other economic sectors as well as all the people living in Canada, Mexico, and the United States. Second, state and capital agents, citing secrecy and expediency, deliberately excluded all other society members, particularly any opposing forces.

The neo-liberal telecommunications changes were continuously con-tested and strongly opposed by the majority of Canadians, including labour, consumer organizations, and public-interest advocates repre-senting the anti-poverty organizations, seniors, and those living in the rural areas of the country. These organizations fought against compe-tition in public telecommunications services (long-distance and basic), the pro-business telecommunications legislation, the lack of public accountability through neo-regulation, the establishment of multi-national rights in two continental trade agreements, and warned about the massive job losses that would occur in the telecommunications sector. Once competition was introduced, these organizations formed broader alliances and coalitions, including even more members of Canadian society who were opposed to these changes. These class and social political actors continued to oppose the dramatic basic local rate increases that followed competition. In addition, they attempted to introduce a narrow subsidy progam to prevent the poor from dropping off the telecommunications network.

Businesses formed broad domestic alliances as well within Canada and the United States. Alliance membership was primarily Canadian and United States multinationals, which benefited from a complex web that included cross-membership in sectoral lobby organizations as well as cross-directorships in the multinational corporations. These alliances were extended to include large telecommunications-user organizations, among others. The purpose of these general and specific capital lobby organizations was to produce consent within their membership, the state, and the general public for the neo-liberal policy approach. Extensive strategies were developed and an inordinate amount of capital and human resources used to fund private research institutes, hold conferences, produce policy position papers, participate in public telecom policy hearings, and appear before legislative standing committees. The view put forward by the various capital lobby organizations included the benefits of liberalization, neo-regulation, privatization, and continental free trade and investment.

The Canadian government in particular played an important role in the production of consent domestically, continentally, and internationally. Moreover, the federal government attempted to produce wide consent for the project by holding a public royal commission on the country's economic prospects. This first effort at building consent only succeeded in heightening citizen opposition. Other federal strategies included the involvement of specific major government departments, such as Foreign Affairs and International Trade, Industry Canada, and the former Department of Communications, to advance neo-liberal policies and continental free trade. A subdepartment within Foreign Affairs and International Trade, the International Trade Advisory Committee, with a number of sectoral advisory groups, was created with the sole purpose of producing consent for continental free trade. The previous resistance to free trade was silenced by state bureaucrats, who invited mainly the chief executive officers of businesses and the presidents of the business lobby organizations; the only other participants were the few token labour representatives. Negotiating both continental trade agreements proved to be even more exclusive. Only Canadian, Mexican, and u.s. bureaucrats and multinational business leaders participated in formulating agreements that affected so many people in all three countries. Shrouded in secrecy and political intrigue, this negotiating process provided all three states with another important vehicle to silence opposing voices.

As the neo-liberal telecommunications approach developed further, the Canadian state became creative in the ways it silenced domestic opposition. By setting a competitive public policy agenda and rapidly advancing the number of hearings dealing with competitive and regulatory reform, the federal communications regulator put more unrealistic time constraints and economic burdens on the groups and organizations fighting to preserve the social aspects of public policy. Throughout the 1990s many public telecom decisions have been written up as if competition was not opposed by trade unions, consumer organizations, anti-poverty organizations, or seniors, among others. The greatest attempt to silence and confuse resistance occurred as a result of a specific federal strategy to keep the opposing groups divided. This strategy was used not only by the federal cabinet and departments but by the federal communications regulator, and with some success, as one of the opposing groups, the Consumers' Association of Canada, was co-opted and exploited. Blending the telecommunications operations of the Department of Communication and the Department of Consumer and Corporate Affairs into a superministry not only reduced the size of the state; it also removed or narrowed entry points, or the means for labour, public-interest, and consumer groups to air their concerns.

Although the Canadian government has gone to great lengths to silence resistance to the new telecom environment, it has, to date, not succeeded in producing complete silence. Contradictions continue to surface regarding the imagined competition in the telecom service sector, the new forms of regulation, quality of service, further basic local rate increases, price increases for customers in rural and remote areas, and increases in the number of people dropping off the system. Uncertainty about whether consent for policy changes will continue from communications businesses, particularly if the major competitors in the cable or telephone industry experience market failure, also contributes to the instability of the sector.

Whether there will be stability for Canadian telecommunications in a continental environment is still an open question. Until recently, the telecommunications service sector has benefited from the fact that no serious market failure has occurred and that consumer groups and public-interest advocates have not formed any cross-border or international coalitions targeting telecom service issues. Telecommunications labour organizations, however, have formed continental alliances and have begun to hold continental conferences at which information is shared among the unions in all three countries in order to fight downsizing and wage and benefit losses. In addition, these unions are involved in aggressive organizing campaigns, not only within North America but internationally as well. Some of the unions continue to build alliances with the progressive community, consumer and public-interest organizations. Whether and to what effect domestic, continental, or international conflicts will contribute to an unravelling of the North America continental regime, only time and further research will tell.

The continentalization of Canadian telecommunications, North American "free trade," and multinational corporations rely on domestic neo-liberal policy reform and the principle of global market expansion. A decade ago NAFTA and CUSFTA, combined with domestic regulatory reform, helped to pave the way for liberalization, neo-regulation, privatization, and internationalization. The WTO, in preparation for increased trade in telecom and multimedia services offered by converged-merged communications, media, and electronic networks, has recommended that member countries adopt the neo-liberal policy model towards international communications and the media. To participate in world trade in multimedia, computer, and telecommunications services, as well as in cultural industries, now requires accepting neo-liberal ideology and policy reforms that increasingly have major implications for democracy and sovereignty. It is far from certain what consequences the neo-liberal experiment will produce. That is why further research that focuses on the social impact of global multimedia and communication networks will remain so critically important for some years to come.

Appendices

Appendix One

Abitibi Price Inc.
Alberta Energy Co.
Alcan Aluminum Ltd.
Algoma Steel Corp. Ltd.
Allied Signal Canada Ltd.
American Express Canada Ltd.
Atco Ltd.
Bank of British Columbia
Bank of Montreal
Bank of Nova Scotia
Peter Bawden Drilling Ltd.
Bechtel Canada Engineers Ltd.
Bell Canada Enterprises Inc.
Benson & Hedges (Canada) Inc.
Bow Valley Industries Ltd.
BP Canada Inc.
Brascan Ltd.
BC Forest Products Ltd.
BC Resources Investment
BC Telephone Company
Burns Foods Ltd
Burns Fry Ltd.
CAE Industries Ltd.
Canada Cement Lafarge Ltd.
Canada Life Assurance Co.
Canada Packers Inc.
Canada Wire and Cable Ltd.
Canadian Chamber of Commerce
Canadian Corporate Management Co.
Canadian Depository for Securities
Canadian General Electric Ltd.
Canadian Imperial Bank of Commerce
Canadian Liquid Air Ltd.
Canadian Manufacturers' Association
Canadian Pacific Ltd.
Canadian Superior Oil Ltd.
Canadian Tire Corp. Ltd.
Canfor Corporation

Canron Inc.
Cargill Ltd.
Cassels Blaikie & Company Ltd.
Celanese Canada Ltd.
Chieftan Development Co.
CIL Inc.
CIP Inc.
Commonwealth Construction Co.
Conseil du Patronat du Quebec
Consolidated Natural Gas Ltd.
Consumers' Gas Co.
Continental Bank of Canada
Continental Cam Canada Ltd.
Cooper Canada Ltd.
Coopers & Lybrand
Co-Steel International Ltd.
Crown Life Insurance
CSL Group Ltd.
Deloitte Haskins & Sells
Dofasco Inc.
Dominion Securities Ltd.
Dominion Textile Inc.
Domtar Inc.
Du Pont Canada Inc.
ERCO
Ethyl Canada Ltd.
Federal Industries Ltd.
Flec Manufacturing Inc.
Ford Motor Company
Fraser Inc.
General Foods Inc.
Genstar Corp.
General Mills Canada Inc.
Goodyear Canada Inc.
Gulf Canada Ltd.
Hedwyn Communications Inc.
Honeywell Inc.
IBM Canada Ltd.

Imasco Ltd.
Imperial Oil Ltd.
Inco Ltd.
Inter-City Gas Corp.
Interprovincial Pipe Line Ltd.
The Investors Group
IPSCO Inc.
ITT Canada Ltd.
Kodak Canada Inc.
John Labatt Ltd.
Laurential Group Corp.
3M Canada Inc.
MacMillan Bloedel Ltd.
Manalta Coal Ltd.
Manufacturers Life Insurance
Marsh & McLennan Ltd.
McLeod Young & Weir Ltd.
William M. Mercer Ltd.
Merrill Lynch Canada Inc.
Metropolitan Insurance Co.
Mollenhauer Ltd.
Molson Companies Ltd.
Montreal City & District Savings Bank
Mutual Life Assurance of Canada
Nabisco Brands Ltd.
National Bank of Canada
NB Telephone Ltd.
Ontario Paper Co. Ltd.
Noranda Inc.
Norcen Energy Resources Ltd.
Northern Telecom Ltd.
Ontario Paper Co. Ltd.
Oshawa Group Ltd.
Placer Development Ltd.
Power Corporation of Canada
Provigo Inc.
Prudasco Inc.
Quaker Oats of Canada
RCA Inc.

Source: BCNI *Annual Report* 1986.

Appendix Two

BCNI EXECUTIVE AND POLICY COMMITTEE 1988, 1994

Enterprise & Head Office	SIC* Code	Type of Business**	1988	1994
EXECUTIVE				
BCNI (2 members)		L	x	x
Canfor Corp.	m	M	PC	x
Canadian Pacific Ltd.	t/c	M	PC	x
Dominion Textiles	t	M		x
Nova Corp.	m	M		x
POLICY COMMITTEE				
Abitibi-Price	m	M	x	
Alcan Ltd.	m	M		x
Bank of Montreal	f	M		x
Bell Canada Enterprises	c	M	x	x
Burns Fry Ltd.	m/w	M	x	
Cargill Ltd.	m	M		x
CIBC	f	M		x
CIL Ltd.	m	M		x
Co-Steel Inc.	m	M	x	
Domtar	m	M	x	
Dupont	m	M		x
DMR Group	s			x
Federal Industries Ltd.	t/w	C	x	
Ford of Canada	m	M	x	
Fortis Inc.	u	C		x
Ganong Bros. Ltd.	m	C		x
General Electric	m	M	x	
Guillevin International	m	M		x
IBM	m	M		x
Imasco Ltd.	m	M		x
ITT Canada	m	M		x
Laurentian Group	f	M	x	
Loran Corp. (Exxon)	r	M		x
Maclean Hunter	c	M	x	x
Manalta Coal Ltd.	r		x	
Molson Co. Ltd.	m	M		x
National Bank of Canada	f	M	x	x
Noranda Inc.	r	M		x
Northern Telecom Ltd.	c/m	M		x

Enterprise & Head Office	SIC* Code	Type of Business**	1988	1994
PanCanadian Petroleum	r	M		x
Price Waterhouse	s			x
Revalstoke Co.	r	M	x	
Royal Bank of Canada	f	M	x	
Sears Canada Ltd.	w	M	x	
Shell Canada	r	M	x	
Stelco Inc.	m	M	x	x
Sun Life Assurance	f	M		x
Trans Alta Utilities	u	M		x
Westcoast Energy Inc.	r	M		x
Wood Gundy	s		x	
Canadian Chamber of Commerce		L	x	x
Canadian Manufacturers' Assn.		L	x	x
Counseil du Patronat du Québec		L	x	x

Source: BCNI, *Annual Report*, 1988, 1994/5.

Legend:

 * c = communication,, f = finance, m = manufacturing, r = resource, s = service, t = transportation, u = utilities, w = wholesale

** L = Lobby, M = multinational, C = corporation

Appendix Three

BCNI FINANCE MEMBERSHIP*

Business	Finance Code	1986	1994
Bank of British Columbia	B	x	
Bank of Montreal	B	x	x
Bank of Montreal Investment Council	Iv		x
Bank of Nova Scotia	B	x	x
Canadian Life Assurance	I	x	x
Canadian Imperial Bank of Commerce	B	x	x
Continental Bank of Canada	B	x	x
Crown Life Insurance	I	x	x
Culinar Inc.**	I	x	x
Great West Life Assurance	I		x
Hongkong Bank of Canada	B		x
Investors Group (Power Corp. 74%)	Iv	x	x
Laurentian Bank of Canada (Group)	B/Iv	x	x
London Insurance Group (Trilon Corp. 56%)	I		x
Manufacturers Life Assurance	I	x	x
McLeod Young and Weir	Iv	x	
Montreal City and District Savings Bank	B	x	
Mutual Life Assurance Company of Canada	I	x	x
National Bank of Canada	B	x	x
National Trust Company	T	x	
Nesbit Burns Inc. (Bank of Montreal)	B		x
Power Corporation	I/Iv	x	x
Prudential Assurance Company Ltd.	I		x
Royal Bank of Canada	B	x	x
Royal Insurance Company of Canada	I	x	x
Royal Trust Company (Royal Bank)	T	x	
Scotia McLeod Inc. (Bank of Nova Scotia)	Iv		x
Standard Life Assurance Company	I	x	x
Sun Life Assurance Company of Canada	I	x	x
Toronto-Dominion Bank	B	x	x
Traders Group Ltd	Iv	x	
Travelers Canada	I	x	

Source: BCNI, *Annual Report*, 1986, 1994.

Legend

* Finance = Banking (B), Insurance (I), Investment (Iv), Other (O).

** Culinar Inc. (Government of Quebec, Confédération des Caisses Populaires et d'Économie Desjardins

Appendix Four

Alcan Aluminium Ltd.
BC Tel
BCE Inc.
Bank of Montreal
Bank of Nova Scotia
Bell Canada
Bombardier Inc.
Brascan Ltd.
Burns Fry Ltd.
CAE Inc.
La Caisse de dépot et placement
    du Quebec
Canada Life Assurance
Canada Trust
Canadian Petroleum Producers
Canadian Bankers' Association
Canadian Chamber of Commerce
Canadian Federation of Independent Business
Canadian Imperial Bank of Commerce
Canadian National
Canadian Pacific Ltd.
Canadian Pulp & and Paper
    Association
Ciba-Geigy Canada Ltd.
Consumers Gas
Coopers & Lybrand
Co-Steel Inc.
Crown Life Insurance Co. Ltd.
Thomas d' Aquino (BCNI)
Deliotte & Touche
Dominion of Canada General
    Insurance Co.
Empire Life Insurance Co.
Export Development Corp.
Ford Motor Company of Canada
Four Seasons Hotel
Great-West Life Assurance Co.
Hewlett-Packard (Canada)
Hydro-Quebec

IBM Canada Ltd.
Imasco Limited
Imperial Oil Ltd.
Inco Ltd.
IPSCO Inc.
London Life Insurance Co.
Maclean Hunter Ltd.
MacMillan Bloedel
Manufacturers Life Insurance Co.
MT&T
Molson Companies Ltd.
Montreal Trust
Mutual Life Assurance (Canada)
National Trust
Noranda Inc.
NOVA Corp. of Alberta
Ontario Hydro
Power Corporation of Canada
Price Waterhouse
Richardson Greenshields of Canada
Royal Bank of Canada
Shell Canada Ltd.
Southam Inc.
Standard Life Assurance Company
Sun Life Assurance of Canada
Suncor Inc.
TELUS Corp.
3 M Canada Inc.
Toronto Dominion Bank
Toronto Stock Exchange
TransCanada Pipelines Ltd.
Unilever Canada Ltd.
Via Rail Canada Ltd.
Westcoast Energy Inc.
Wood Gundy Inc.
Xerox Canada

Source: C.D. Howe Institute,
The NAFTA Policy Study 21, 1994.

Appendix Five

PARTIAL LIST OF BOARD OF TRUSTEES OF THE FRASER INSTITUTE, 1994

Argus Corporation
BCA Publications Limited
Bank of Nova Scotia
Bata Limited
CAE Industries
Canadian General-Tower Ltd
Canadian Hunter Exploration
Canadian Imperial Bank of Commerce
Canadian Pacific Limited
Candor Investments
Delcan Corporation

Hollinger Inc.
Inwest Developments
IPSCO Inc.
Loblaw Companies Ltd.
MacMillan Bloedel Ltd.
Noranda Mines
Prospero Realty International Inc.
Royal Bank of Canada
University of Manitoba

Source: Fraser Institute, 1994 Board of Trustees.

Appendix Six

PARTIAL LIST OF BRIEFS
AND INTERVENTIONS FOR
CHALLENGES AND CHOICES
(MACDONALD COMMISSION), 1984

Action travail des femmes
  du Québec
Agriculture Institute of Canada
Alliance of Canadian Cinema,
  Television and Radio Artists
Anglican Church of Canada
Asia Pacific Foundation of Canada
Association of Canadian Distillers
Association of Canadian Financial Corp.
Associations of Universities
  & Colleges of Canada
Atomic Energy of Canada
Bank of Montreal
Bell Canada
Bell Canada Enterprises Inc.
BC Federation of Women
Burns Food Ltd.
Burns Fry Ltd.
BCNI
C.D. Howe Institute
CNCP Telecommunications
Canada West Foundation
Canadian Advisory Council
  on the Status of Women
Canadian Association of University Teachers
Canadian Business and Industry
International Advisory Council
Canadian Business Equipment
  Manufacturers Association
Canadian Chamber of Commerce
Canadian Chemical Producers
  Association
Canadian Conference of Catholic Bishops
Canadian Export Associations
Canadian Federation of Agriculture
Canadian Federation of Communication Workers

Canadian Federation of Independent Business
Canadian Federation of Labour
Canadian Federation of Students
Canadian Gas Association
Canadian General Electric
Canadian Hospital Association
Canadian Labour Congress
Canadian Manufacturers' Association
Canadian Pensioners Concerned Inc.
Canadian Teachers Federation
Canadian Textiles Institute
Canadian Union of Public
    Employees
Celanese Canada
Communications Workers of Canada
Confédération des caisses
    populaires et d'économic
    Desjardins du Québec
Council for Yukon Indians
Council of National Ethnocultural
Organizations of Canada
Dene Nation of the NWT
Dominion Textile Inc.
Dow Chemical Canada
Economic Council of Canada
Environment Canada
Falconbridge Ltd.
Federated Anti-Poverty Groups
    of BC
Federation of Canadian
    Municipalities
Governments of Alberta, Manitoba, Sask., Quebec,
    NB, NS, PEI, Nfld.,Yukon, NWT
Gulf Canada Ltd.
NB Telephone Co.
North-South Institute
Northern Telecom Ltd.
Petro-Canada
Placer Development Ltd.
Public Interest Advocacy Centre
Science Council of Canada
SNC Group

Social Action Commission,
  Roman Catholic Diocese
Ultramer Canada
United Church of Canada, Working Unit on Social Issues & Justice
United Mine Workers of America
United Steelworkers of America
Vanier Institute of the Family
Via Rial Canada Inc.

Source: *Challenges and Choices*

Appendix Seven

PARTICIPANTS IN *THE OTHER MACDONALD REPORT*, 1985

Canadian Conference of Catholic
  Bishops
Canadian Institute for Economic
  Policy
Canadian Mental Health Association
Canadian Union of Public Employ-
  ees
Confederation of National Trade
  Unions
Makivik Corporation
Manitoba Political Economy Group
National Action Committee on the
  Status of Women
National Council of Welfare
National Farmers Union
Ontario Public Service Employees
  Union
Quebec Teachers Federation
Social Planning Council of Metro-
  politan Toronto
United (Canadian) Auto Workers
United Church of Canada
Vanier Institute of the Family
Women Against the Budget

Source: Drache and Cameron, eds., *The Other Macdonald Report*, 1985.

Appendix Eight

SELECTED MEMBERS OF THE INSTITUTE FOR RESEARCH ON PUBLIC POLICY, 1982

BOARD OF DIRECTORS

The Honourable John Aird
Bank of Nova Scotia
Burns Fry Ltd.
Chamber of Commerce (Calgary)
Employers' Council of British Columbia
Laurentian Fund Inc.
McCarthy & McCarthy
McCuaig, Desrochers
Molson Companies
Newfoundland Fishermen, Food and Allied Workers' Union
University of New Brunswick (Finance & Administration)
University of Montreal (Department of Economics)

EXECUTIVE COMMITTEE

The Honourable Robert Stanfield
Louis Desrochers

INVESTMENT COMMITTEE

The Honourable Donald Macdonald (Macdonald Commission)
Tom Kierans

MEMBERS AT LARGE

Council of Maritime Premiers
Dalhousie University
Economic Council of Canada
Osgoode Hall Law School
St Mary's University
Science Council of Canada
Status of Women Canada
Telesat Canada
Universities and Colleges of Canada
University of Calgary
University of Laval
University of Manitoba
University of Montreal
University of Ottawa
University of Toronto
University of Saskatchewan

Source: *Regulatory Reform in Canada*, 1982: 131–3.

Appendix Nine

CNCP TELECOMMUNICATIONS: INTERCONNECTION WITH BELL CANADA PUBLIC HEARING PARTICIPANTS, FEBRUARY 1978 (CRCT 79–11)

TELECOMMUNICATION
CARRIER AGENTS

Alberta Government Telephone
The Atlantic Telcos (Newfoundland
  Tel, Island Tel, Maritime
  Telephone & Telegraph,
  New Brunswick Tel)
Bell Canada
BC Telephone
CNCP Telecommunications
Manitoba Telephone
Saskatchewan Telephone

UNION AGENTS

Canadian Federation of Communi-
  cation Workers
Canadian Railway Labour
  Association

STATE AGENTS

CRTC
The Director of Investigation &
  Research, Combines Investigation
  Act
Government of British Columbia
Government of Ontario
Government of Quebec
Governments of Atlantic Provinces

BUSINESS LOBBY AGENTS

Canadian Airlines Telecommunications Association

Canadian Association of Broadcasters
Canadian Association of Data
  Processing Service Organizations
Canadian Business Equipment
  Manufacturers Association
Canadian Cable Television
  Association
Canadian Federation of Indepen-
  dent Business
Canadian Information Processing
  Society
Canadian Industrial Communica-
  tions Assembly Business Interve-
  nors Society of Alberta
Canadian Manufacturers' Associa-
  tion
Canadian Petroleum Association
Canadian Press
Canadian Trucking Association
National Association of Canadian
  Credit Unions
Pharmaceutical Manufacturers
  Association of Canada
Retail Council of Canada

BUSINESS AGENTS-TELECOM
USERS & EQUIPMENT
MANUFACTURERS

Aquils BST Ltd.
Bank of Nova Scotia
Banque Canadienne Nationale
la Banque d'Epargne de la Cité
  et du District de Montréal
Calgary Chamber of Commerce
Canadian General Electric Ltd.

Comshare Ltd.
Control Data Canada Ltd.
Domtar Ltd.
Durcos
Fédération des Caisses s'entraide
  économique du Québec
IBM Canada Ltd.
Industrial Life-Technical Service Inc.
Interprovincial Pipelines Ltd.
Informaticiens Associés de Montréal
  Inc.
la coopérative fédérée de Québec Inc.
Les Auto-Parts Reliés par Télex
  Meilleur
Métro-Richelieu Inc.

Multipal Access Ltd.
National Data Co.
Provigo Inc.
Richardson Securities of Canada
  Ltd.
Roy & Associés
Royal Bank of Canada
Simpson-Sears Ltd.
Systems Dominions Ltd.
Traders Group Ltd.
Westinghouse Canada Ltd.

Source: CRTC Decision 79–11, CNCP Telecommunications: Interconnection with
Bell Canada, May 1979.

Appendix Ten

TELECOM DECISION CRTC 85–19: INTEREXCHANGE COMPETITION AND RELATED ISSUES, PARTICIPANTS

TELECOM CANADA MEMBERS

Alberta Government Telephone
  (AGT)
Bell Canada
British Columbia Telephone (BC Tel)
Maritime Telephone & Telegraph
  (MT & T and Island Telephone)
Manitoba Telephone Systems (MTS)
New Brunswick Telephone (NB Tel)
Newfoundland Telephone (NewTel)
Telesat

CHALLENGERS

BC Rail
CNCP Telecommunications

OTHER TELEPHONE COMPANIES

Edmonton Telephone
Hurontario Telephone Ltd.

OTHER COMMUNICATION
COMPETITORS

CMQ Communications Inc.
NorthwesTel
Telephone Answer Association
  of Canada
Tera Nova

USER ORGANIZATIONS

Association of Competitive Telecommunications Suppliers
Broadcast News Ltd.

Canadian Bankers Association
Canadian Business Equipment
  Manufacturers Association
Canadian Business Telecommunica-
  tions Alliance (CBTA)
Canadian Cable Television Association
Canadian Independent Telephone
  Association
Canadian Petroleum Association
Canadian Press
Canadian Radio Common Carriers
  Association
Cantel Cellular Radio Group
CTG Inc.
Hotel Association of Canada Inc.
Ontario Hotel and Motel Associa-
  tion (OHMA)

GOVERNMENT ORGANIZATIONS

Director of Investigation and
  Research, Combines Investigation
  Act the Director)
Government of Ontario
Government of Quebec
Ontario Municipal Electric Associa-
  tion (OMEA)

UNIONS

Canadian Federation of Communi-
  cation Workers (CFCW)
Communications, Electronic,
  Electrical, Technical & Salaried
  Workers of Canada (CWC)
Telecommunications Workers' Union
  (TWU)

PUBLIC INTEREST AND
CONSUMER GROUPS

Consumers' Association of Canada
(CAC)
Federated Anti-Poverty Groups of
B.C., B.C. Old Age Pensioners'
Organization, Kennedy House
Senior Recreation Centre, Lower
Mainland Alliance of Informa-
tion Referral Services, Save Our
Shores, Society for Promoting
Environmental Conservation,
the British Columbia Provincial
Council of Women, United Elders'
Association of British Columbia,
and West End Seniors Network
(FAPG et al)
National Anti-Poverty Organization
(NAPO)

CRTC COMMISSIONERS

John Lawrence (Chairman to May
1985)
André Bureau
Monique Coupal
Paul H. Klingle
Jean-Pierre Mongeau

Source: CRTC Decision 85–19, *Interexchange Competition and Related Issues*,
Aug. 1985: 4, 5.

Appendix Eleven

CANADIANS FOR COMPETITIVE
TELECOMMUNICATIONS
MEMBERSHIP LIST, 1986

Association of Competitive Telecommunications Suppliers
Canadian Association of Data & Professional Services Organizations
Canadian Bankers' Association
Canadian Business Equipment Manufacturers Association (CBEMA)
Canadian Business Telecommunications Alliance (CBTA)
Canadian Federation of Independent Business (CFIB)
Canadian Organization of Small Business
Canadian Radio Common Carriers Association
Hotel Association of Canada
Ontario Hotel and Motel Association

Source: Canadians for Competitive Telecommunications, Feb. 1986.

Appendix Twelve

INFORMATION TECHNOLOGY ASSOCIATION
OF CANADA *ITAC* MEMBERS AS OF 1995

* 3M Canada Inc.
  Acorn Partners
  Agensys (Canada) Ltd.
* Alias Research Inc.
  Alliance for Converging
    Technologies
* Ambrex Technologies Inc.
* Amdahl Canada Ltd.
* Anderson Consulting
  Anne McKague & Associates
* Apple Canada Inc.
  Arnott Design Group Inc.
  Arthur Anderson & Co.
* Atlantic Computer Institute
* AT&T Canada Inc.
* AT&T Global Information
  Solutions Canada Ltd.
* Auto-Trol Technology Ltd.
  A.D. Parker & Associates
  AV International Inc.
  Baker Harris & Partners Ltd.
  BMC Group/Canada
  BDP Business Data Services Ltd.
  Blake, Cassels & Graydon Borden
    & Elliot
* Brant Interprovincial Systems Ltd.
* Bull HN Information Systems Ltd.
  Burwell Information Services
  Callstream Communications Inc.
  Canada NEWSwire Ltd.
  Canada Post Corporation
  Canadian Industrial Innovation
    Centre/Waterloo
  Canadian Music Industry
    Database
  Canadian Standards Association
  Canadian Tire Corporation Ltd.
  Canadore College

  CANARIE
  Cannex Financial Exchanges
  Carswell Electronic Publishing
  CCH Canadian Ltd.
  CDSL
  CEDROM-SNI Inc.
  Centennial College
  Centre d'Innovation en
    Technologies de l'Information
* CGI Group Inc.
  Cisco Systems Canada Ltd.
  City of Fredericton
  Clinicare Corp.
  CM Inc.
* Compaq Canada Inc.
  ComputeChan Computer
    Systems Corp.
* Computer Methods
  Computer Servicenters Ltd.
* Comshare Ltd.
  ComTel Plus Inc.
  Conseillers en Gestion GRV
  Contractors Network Corp.
  Control Data Systems Canada Ltd.
  Corporate Personnel Recruiters Ltd.
  Cray Research (Canada) Inc.
  Crossley Mann
  Dart C.P. Services Ltd.
  Datacor/ISM Atlantic Corp.
* Data General (Canada) Ltd.
  Deeth Williams Wall
  Delphax Systems
  DeVry Institute of Technology
* Digital Equipment of Canada Ltd.
  Ditek Software Corp.
* DMR Group Inc.
  Durnford Management Corp.
  Dymaxion Research Ltd.

* EDS Canada
Effectivation Inc.
* Elan Data Makers
Enterprise Network Inc.
Ernst & Young
Espial Productions Ltd.
* Evans Technologies Inc.
Eyepoint Inc.
Faxfirst Canada Inc.
Fraser & Betty
Fulcrum Technologies Inc.
F/X Corporation
Global Information Services
Globestar Systems Ltd.
G.T.C. Transcontinental Group Ltd.
Harris Donovan Systems Ltd.
* Hewlett-Packard (Canada) Ltd.
Hill and Knowlton Canada Ltd.
* IBM Canada Ltd.
InContext Corporation
Industrial Research &
    Development Institute
Info J.E.D
Information Highways
Information TechnologyResearch
InGenius Engineering Inc.
Inquix Consulting Ltd.
Interactive Image Technologies Inc.
International Data Corp.
    (Canada) Ltd.
* ISM Information Systems
Management Corp.
* IST Computer Service Co.
IT Systems Group
Johnson Lake Software Ltd.
KAO Infosystems Canada Inc.
KBS Technology Inc.
Lang Michener Lawrence & Shaw
LANServe Technologies Inc.
* LGS (Group) Inc.
Lightstone & Associates
* Linktek Corporation
* Lotus Development Canada Ltd.

Lyons Technology Group Inc.
Management Dimensions Ltd.
Management Services Corp.
Manufacturing Concepts
Manulife Financial
Marcia Olmsted & Associates
Maritime Information
    Technology Inc.
McCarthy Tetrault
McGraw-Hill Ryerson Ltd.
Medco Data Systems Ltd.
Medisoft Inc.
Megalith Technologies Inc.
* Metcan Information
    Technologies Inc.
Micromedia Ltd.
* Microsoft Canada Ltd.
Miller Thomson
Mildred and Trebilcock
Minicom Data Corp.
Mortice Kern Systems Inc.
MPACT Immedia
Multi-Health Systems
Neill and Gunter Ltd.
Network Data Systems Ltd.
New Paradigm Learning Corp.
Newcourt Credit Group Inc.
Newton Associates
* Northern Telecom Canada Ltd.
Novatec Systems
Numetrix Ltd.
Oak Technologies Inc.
* Oracle Corp. Canada
ORTECH Corporation
OTEC Software Corp.
P & P Data System Inc.
Peat Marwick Thorne,
* Pitney Bowes of Canada Ltd.
* PMP Associates Inc.
Proactive Technology
    and Trading Prodigm Inc.
PSC, The Public Sector
    Corporation

Pythonic Management Corp.
Research in Motion
Riley Informaiton Services Inc.
* Rogers Communications Inc.
Ross Hutchison & Associates Inc.
Rust Associates
Ryerson Polytechnic University
R. J. Pritchard & Associates
S & S Software Ltd.
SAP Canada Inc.
SAS Institute (Canada) Inc.
Servacom
Services Documentaire Multime-
    dia Inc.
Sextant Software
* SHL Systemhouse
* Siemens Electric Ltd.
* Sierra Systems Consultants Inc.
* Silicon Graphics Canada Inc.
Sim*Pax Systems Inc.
Sim, Hughes, Ashton & McKay
SIRIT Technologies Inc.
Skillbase Resources Ltd.
Smith, Lyons, Torrance,
Stevenson & Mayer
SMW Advertising
SoftCapital Corp. Development Inc.
* Software Brokers Ltd.
Software Human Resources
    Council
* Softworld
* Southam City & Community
    Newspapers
Southam Information
    & Technology Group

* Spar Aerospace Ltd.
Sprint Canada
Stanley L. Jacobs Research Ltd.
Star Data Systems Inc.
* Stentor Canadian Network
    Management
* Stentor Resource Centre Inc.
* Stentor Telecom Policy Inc.
Storm Technical Communica-
    tions Inc.
Strategic Decisions Inc.
Stratus Computer Corp.
* Sun Microsystems of Canada Inc.
Sunnybrook Health Science
    Centre
Switchview Inc.
Sykes Enterprises Inc. of Canada
* Tandem Computer Canada Ltd.
Tektronix Canada Inc.
Teleglobe Canada Inc.
Telezone Corp.
Teranet Land Information
    Services Inc.
* Unisys Canada Ltd.
* Unitel Communications Inc.
University of Windsor
Videoway Communications
Virtual Corporation
* Wang Canada Ltd.
Ward Associates
Watcom Corporation
Webcom Technical Documentation
* Xerox Canada Inc.

Source: ITAC Members List and *Annual Reports*, 1993, 1994, 1995.
* Board of directors of ITAC.

Appendix Thirteen

CANADIAN BUSINESS
TELECOMMUNICATIONS ALLIANCE (CBTA):
MEMBERS AS OF 1991

COMPANIES & CROWN
CORPORATIONS

3M Canada Inc.
Abitibi-Price Inc.
Account-a-Call Canada
Aetna Life Insurance Co.
Agropur Co-Op, Agro-Alimen
Air Canada
Alberta Energy Co. Ltd.
Alcan Aluminum Ltd.
Allstate Insurance
Amdahl Canada Ltd.
American Express Canada Inc.
Amoco Canada Petroleum Co. Ltd.
Answer Plus Corporation
Aviscar Inc.
BC Ferry Corporation
BC Gas Inc.
BC Hydro
BC Lottery Corporation
BC Systems Corporation*+
Bank of Canada
Bank of Montreal+
Bank of Nova Scotia~
Banque Nationale du Canada
Baxter Corporation
Bell & Howell Canada Ltd.
Blackburn Media Group*
Block Bros. Industries Ltd.
Bull Hein Info. Systems Ltd.
Burns Fry Ltd.
Call-Net Telecommunications
Campbell Soup Company Ltd.
Canada Life Assurance Co.
Canada Mortgage & Housing Corp.
Canada Newswire Ltd.

Canada Packers Inc.
Canada Post Corp.
Canada Steamship Lines Inc.
Canada Trust Co.
Canada Wire & Cable Ltd.
Canadair Ltd.
Canada ADP Services Ltd.
Canadian Airlines Int'l
Canadian Bankers Association+
Canadian Broadcasters Corporation
Canadian Depository/Securities
Canadian General Electric Co.
Canadian General Insurance
Canadian Healthcare Telematics
Canadian Hunter Exploration
Canadian Imperial Bank
   of Commerce+
Canadian National Railways (CNR)
Canadian Pacific Hotels/Resort
Canadian Pacific Ltd.
Canadian Red Cross Society
Canadian Satellite Comm. Inc.
Canadian Tire Acceptance Ltd.
Canadian Tire Corp.#
Canadian Utilities Ltd.
Canadian Wester Natural Gas
Cancom Satlink Business Service*
Canon Canada Inc.
Cargill Ltd.
CCL Industries Inc.
Celanese Canada Inc.
Centra Gas
Central Guaranty Trust Co.
Chedoke-McMaster Hospitals
CIBA-Geigy Canada Ltd.
CIBC Investment Bank
Citibank Canada

Co-Operators Data Services
Co-Operators Financial Services
Coca-Cola Beverages
Cognos Inc.
Comark Services Inc.
Comdisco Disaster Recovery Service
Comdisco Ltd.~
Confederation Life Insurance
Consumers Gas Co. Ltd.*
Consumers Glass
Continental Canada
Crane Canada Ltd.
Crown Life Insurance Co.
Crowntek Business Centre Inc.
Cue Datawest
Datacor Atlantic Inc.
Dataline Inc.
Deere & Co.
Delta Hotels
Digital Equipment of Canada+
Dofasco Inc.
Dominion Textile Inc.
Dow Chemical Canada Inc.
Du Pont Canada Inc.
Dun & Bradstreet Canada Ltd.
Dylex Ltd.
EDS of Canada Inc.*+
Equifax Canada
Ernst & Young
Euro Brokers Canada Ltd.
Export Development Corp.
Falconbridge Ltd.
Fax People (The)
Finning Ltd.
First City Trust Co.
Foothills Hospital
Gandalf Technologies Inc.
Gaz Metropolitan
Gemini Group (The)
General Datacom Ltd.
Glaxo Canada Inc.
Goldman Sachs Canada
Goodyear Canada Inc.

Grace Hospital, CGS Site
Grand & Toy Ltd.
Great-West Life Assurance Co.
GTL Transport (1989) Inc.
Hewlett-Packard Canada Ltd.*+
Hibernia Management
   & Development
Hoechst Canada Inc.
Home Oil Co. Ltd.
Honeywell Ltd.
Hongkong Bank of Canada
Hospital for Sick Children
Household Financial Corp. Ltd.
Hudon et Deaudelin Ltee.
Hudson's Bay Co.
Human Resources Secretariat
Husky Injection Molding System
IBM Canada Ltd.~#+
ICI Canada Inc.
IIS Technologies
Imperial Oil Ltd.
Inco Limited
Indal Computer Services
Infonet Computer Sciences Canada
Infotron Canada Ltd.
Insurance Corp. BC IST*
Janssen Pahrmaceutica
John Labatt Ltd.
Kimberly-Clark Canada Inc.
Kodak Canada Inc.
Kraft General Foods Canada Inc.
LDN Communications
Levesque Beaubien Inc.
Levi Strauss & Co. Canada Inc.
Leviton Telecom
Livingston International Inc.
Loto Quebec
MacKenzie Financial Corp.
Maclean Hunter Limited
Manufacturers Life Insurance
Manulife Financial Canadian
   Maritime Life Assurance Co.~
Matsushita Electric of Canada

McMillan, Binch
Merck Frosst Canada Inc.
Merrill Lynch Canada Inc.
Milgram & Company Ltd.
Mitel Corporation
Mobile Oil of Canada
Montreal Trust
Motorola Canada Ltd.
Motorola Information Systems
Mutual Life of Canada
Nabisco Brands Ltd.
National Arts Centre
National Data Corporation
National Research Council
   of Canada
National Sea Products
National Trust
NEC Canada Inc.
Nesbit Burns+
Nesbitt Thompson Inc.
Nestle Enterprises Ltd.
Newbridge Networks Corp.
Newfoundland & Labrador Hydro
Newfoundland Light & Power Co.
Noranda Inc.
Norex Leasing Inc.
North American Life
Northern Telecom Canada Ltd.#
Nova Corporation of Alberta
Nova Scotia Power Corporation*
Novacor Chemicals (Canada) Ltd.
Ogilvy, Renault
Olympia & York Developments Ltd.
Ontario Hospital Association/Blue
   Cross
Ontario Hydro#+
Ortho Pharmaceutical Canada
Oshawa Group
Peat Marwick Stevenson Kellogg
Peat Marwick Thorne
Peel Memorial Hospital
Petro-Canada Inc.
PHH Canada Inc.

Phillips Electronics Ltd.
Pillsbury Canada Ltd.
Pitney Bowes Canada Ltd.
Polysar Rubber Corp.
Pratt & Whitney Canada Inc.*
Prenor Trust Co. of Canada
Procor Ltd.*
Provigo Inc.*
Prudential Assurance Co. Ltd.
Prudential Insurance Co. of America
Purolator Courier Ltd.
Quaker Oats Co. of Canada
Reader's Digest Association
   of Canada
Reckitt & Colman Canada Inc.
Regie L'Assurance Auto de Quebec
Richardson & Greenshields
Ricoh Canada Ltd.
Royal Bank of Canada Ltd.~*##+
Royal Insurance Co. of Canada
Royal Trust
Royal Victoria Hospital
S.C. Johnson and Son Ltd.
SITA
Safety Supply Canada Ltd.
Saskatchewan Property Manage-
   ment Corp.
Science North
ScotiaMcleod Inc.
Scott Paper Ltd.
Seaboard Life Insurance Co.
Sears Canada Inc.
Shell Canada Ltd.
Sherritt Gordon Ltd.
SHL Systemhouse Inc.*
Sidbec-Dosco Inc.
Societe GRICS
Southam Business Information
Southam Inc.*+
Spar Aerospace Ltd.
St. Michael's Hospital
State Farm Insurance
Steinberg Inc.

Stelco Inc.
STM Systems Corp.*
Storagetek Canada Inc.
Sun life Assurance Co.~
Sunoco Inc.
Syncrude Canada Ltd.
T. Eaton Company
Technologie la Laurentienne
Telerate Canada Inc.
Thomson Newspapers+
TNT Management Service
Toronto Hydro
Toronto Star (The)
Toronto Stock Exchange
Toronto Transit Commission
Toronto-Dominion Bank+
Toyota Canada Inc.
Transalta Utilities Corp.
TransCanada Pipelines
Trust General Canada
Ultramar Canada Inc.
Union Carbide Canada Ltd.
Union Gas Ltd.
Unisys Canada Inc.
Unitel*#
United Co-Operators of Ontario
United Grain Growers Ltd.
United Parcel Service+
Vancouver Stock Exchange
Venture Inns Inc.
VIA Rail Canada Inc.
Victoria Hospital Corp.
Warner-Lambert Canada Inc.
Wellington Insurance
Wester Canada Lottery Corp.
Westinghouse Canada Inc.
Weston Foods Ltd.
William Mercer Ltd.
Woman's College Hospital
Woodward Stores Ltd.
Woolworth Corp.
Workers' Compensation Board
   of BC

Workers' Compensation Board
   of Ontario
Xerox Canada Inc.
York-Finch General Hospital
Zurich Canada

FEDERAL GOVERNMENT
OF CANADA

Department of Communications
Employment & Immigration
Energy, Mines & Resources
Department of Fisheries & Oceans
Revenue Canada, Customs & Excise
Revenue Canada, Taxation
Transport Canada

PROVINCIAL GOVERNMENTS

Alberta*
New Brunswick
NWT+
Nova Scotia
Ontario
Quebec

EDUCATION

Carleton Board of Education
Carleton University
Centennial College
Humber College
McGill University
McMaster University
Memorial University
   of Newfoundland*
Niagara College of Applied Arts
Northern Alberta Institute
   of Technology
Queen's University
Ryerson Polytechnical Institute*
Sheridan College
Simon Fraser University

University of Alberta Hospital
University of British Columbia
University of Calgary
University of Guelph
University of Manitoba*
University of Ottawa
University du Quebec
University of Toronto
University of Waterloo
University of Western Ontario
York University

MUNICIPALITIES

City of Calgary
City of Mississauga

City of Saskatoon
Corp. of the City of Toronto*
Corp. of Town of Markham
Regional Municipality of Peel+
Regional Municipality of York

Legend:

Past presidents
* Board of Directors, 1990–95
~ Executive Council
+ Telecom Public Policy Committee

Source: Canadian Business Telecommunications Alliance, *Roster of Members*, June
1991, and *Annual Reports*, 1990–95.

Appendix Fourteen

COMMUNICATIONS COMPETITION COALITION (CCC): MEMBERSHIP AS OF 1991

Abitibi Price
ADT Canada Inc.
Alliance of Canadian Travel
  Associations
B.A. Seville and Associates
Bourse de Montreal
BP Canada
Bank of Nova Scotia
Bryker Data
Canadian Alarm and Security
Canadian Bankers Association
Canadian Bankers Association
Canadian Manufacturers Association
Canadian Imperial Bank
  of Commerce (CICB)
Canadian Press
Central Guaranty Trust
Comdisco Disaster Recovery Services
EDS of Canada
Federal Industries of Canada
Frederick Transport
Gemini Group
General Motors
Gesco Industries
Great West Life

Grifcold
IBM
Imperial Oil
Indal Computers
Lawson Morrison Group
Manufacturers Life
Midland Walwyn Ltd.
Moore Corporation
National Bank of Canada
Rex Aviation
Royal Bank
Royal Trust
Sears (Canada) Ltd.
Southam Inc.
STM Systems Corp.
Sun Life
Syscor
Toronto Dominion Bank
Valley City Manufacturing

Source: CRTC Telecom Public Notice 1990–73, *Unitel Communications Inc. & B.C. Rail Telecommunications/Lightel Inc. to Provide Public Long Distance Voice Telephone Services and Related Resale & Sharing Issues*, 18 Apr. 1991.

Appendix Fifteen

TELECOM DECISION CRTC 92–12:
COMPETITION IN THE PROVISION
OF PUBLIC LONG-DISTANCE VOICE
TELEPHONE SERVICES AND RELATED
RESALE AND SHARING ISSUES,
PARTICIPANTS

STENTOR MEMBERS
(FORMALLY TELECOM CANADA)

Alberta Government Telephone Ltd.
   (AGT)
British Columbia Telephone (BC Tel)
Bell Canada
Island Telephone (Island Tel)
New Brunswick Telephone (NB Tel)
Newfoundland Telephone (NewTel)
Maritime Telephone & Telegraph
   (MT&T)

CHALLENGERS

BC Rail Telecommunications
Call-Net Telecommunications and
   Lightel Inc. (joint venture BCRL)
Unitel (CNCP Telecommunications
   and Rogers Communications)

OTHER COMMUNICATIONS
COMPANIES AND ASSOCIATIONS

Association des Compagnies
   de Téléphone du Quebec
BCE Mobile Communications Inc.
Canadian Cable Television Associa-
   tion (CCTA)
Canadian Independent Telephone
   Association and the Ontario
   Telephone Association (CITA/OTA)
Edmonton Telephones Corporation
   (Ed Tel)

Fonorola Inc.
NorthwesTel Inc.
Québec Téléphone (Québec Tel)
Telecom Networks International
   Limited of New Zealand (TNI)
U.S. Intelco Networks Inc. (Intelco)

COMMUNICATIONS-USER
ORGANIZATIONS

Canadian Business Telecommunica-
   tions Alliance (CBTA)
Canadian Federation of Indepen-
   dent Business (CFIB)
Communications Competition
   Coalition (CCC)

GOVERNMENT ORGANIZATIONS

City of Toronto
Director of Investigation and
   Research, Bureau of Competition
   Policy (the Director)
Government of Alberta
Government of British Columbia
Government of Manitoba
Government of Ontario
Government of Québec
Government of Saskatchewan

UNIVERSITIES

McMaster University
University of Toronto

PUBLIC-INTEREST AND
CONSUMERS' ORGANIZATIONS

Association des Consommateurs
du Québec Inc.
BC Old Age Pensioners's Organiza-
tion, Council of Senior Citizens'
Organizations West End Seniors'
Network, Senior Citizens' Organi-
zations, West End Seniors'
Network, Senior Citizens' Associ-
ation, Federated Anti-Poverty
Groups of BC, and Local 1–217
IWA Seniors (BCOAPO et al.)
Canadian Association of the Deaf
Consumers' Association of Canada
(CAC)
National Anti-Poverty Organization
and Rural Dignity of Canada
(NAPO/RDC)

UNIONS

Atlantic Communications & Techni-
cal Workers' Union Communica-
tions and Electrical Workers of
Canada (CWC)
Telecommunications Workers' Union
(TWU)

CRTC COMMISSIONERS

Louis R. (Bud) Sherman (Chairman)
Fernard Bélisle
David Colville
Beverly J. Oda
Edward A. Ross (dissented)

Source: CRTC 92–12, *Competition in the
Provision of Public Long Distance Voice
Telephone Services and Related Resale
and Sharing Issues*, 12 June 1992, pp. 5, 6.

Appendix Sixteen

TELECOM DECISION CRTC 94-19: REVIEW OF REGULATORY FRAMEWORK PARTICIPANTS

STENTOR MEMBERS

AGT Limited
British Columbia Telephone
Bell Canada
Island Telephone Co. Ltd.
Maritime Telephone & Telegraph
New Brunswick Telephone
Newfoundland Telephone
Northwestel Inc.
Stentor Resource Centre Inc.

COMPETITORS

Smart Talk Network
Sprint Canada Inc. (formally Call-Net Telecommunications Inc.)
Unitel Communications Inc. (Unitel)
Telecommunications Inter-Cité 2000
Telecommunications Terminal Systems (TTS)

OTHER COMMUNICATIONS COMPANIES AND ASSOCIATIONS

Allarcom Pay Television Ltd.
Canadian Association of Broadcasters (CAB)
Canadian Cable Television Association (CCTA)
Canadian Satellite Communications Inc.
Changing Concepts of Time (CCOT)
Rogers Cable T.V. Ltd. (RCTV)

COMMUNICATIONS-USER ORGANIZATION

Association of Competitive Telecommunications Suppliers
Canadian Business Telecommunications Alliance (CBTA)
Canadian Daily Newspaper Association (CDNA)
Canadian Federation of Independent Business (CFIB)
Canadian Independent Telephone Association (CITA)
Canadian Satellite Users Association (CSUA)
Competitive Telecommunications Association (CTA)

GOVERNMENT ORGANIZATIONS

Director of Investigation and Research, Bureau of Competition Policy (Director)
Government of British Columbia
Government of Ontario
Gouvernement du Québec
University of Toronto

UNIONS

ACTRA
Communications, Energy and Paperworkers of Canada (CEP/SCEP)
Telecommunications Workers' Union (TWU)

PUBLIC INTEREST AND
CONSUMERS ORGANIZATIONS

BC Public Interest Advocacy Centre,
for BC Old Age Pensioners'
Organization, Council of Senior
Citizen's Organization West End
Seniors' Network, Senior Citi-
zen's Association, Federated
Anti-Poverty Groups of B.C.,

Local 1–217 IWA Seniors
(BCOAPO et al.)
Fédération nationale des consomma-
teurs du Québec and ACQ
(FNACQ)
National Anti-Poverty Organization
(NAPO)

Source: CRTC, 94–19, 1994.

Appendix Seventeen

CANADIAN BANKERS ASSOCIATION: PARTIAL MEMBERSHIP AS OF 1994

DOMESTIC MEMBERS

Bank of Montreal (Mbank)
Bank of Nova Scotia
Canadian Imperial Bank
  of Commerce (CICB)
Canadian Western Bank
Citibank of Canada
Hongkong Bank of Canada
National Bank of Canada
Royal Bank of Canada
Toronto Dominion Bank

FOREIGN MEMBERS*

ABN AMRO Bank of Canada
Amex Bank of Canada
Banco Commerciale Italiana
  of Canada
Banca Nazionale del Lavoro
Banco Central Hispano-Canada
Bank Hapoalim (Canada)
Bank Leumi Le-Israel (Canada)
Bank of America Canada
Bank of Boston Canada
Bank of China (Canada)
Bank of East Asia (Canada)
Bank of Tokyo Canada
Bangue Nationale de Paris (Canada)
Barclays Bank of Canada
BT Bank of Canada
Chase Manhattan Bank
Chemical Bank of Canada
Cho Hung Bank of Canada
Citibank Canada
Crédit Lyonnais Canada
Credit Suisse Canada
Dai-Ichi Kangyo Bank (Canada)
Daiwa Bank of Canada

Deutsche Bank of Canada
Dresdner Bank of Canada
Fugi Bank of Canada
Hanil Bank of Canada
Hongkong Bank of Canada
Industrial Bank of Japan (Canada)
International Commercial Bank
  of Cathay (Canada)
Israel Discount Bank of Canada
Korea Exchange Bank of Canada
Mellon Bank of Canada
Mitsubishi Bank of Canada
Morgan Bank of Canada
National Bank of Greece (Canada)
National Westminster Bank
  of Canada
NBD Bank of Canada
Overseas Union Bank of Singapore
  (Canada)
Paribas Bank of Canada
Republic National Bank
  of New York (Canada)
Sakura Bank (Canada)
Sanwa Bank of Canada
Société Générale (Canada)
Sottomayor Bank of Canada
Standard Chartered Bank of Canada
State Bank of India (Canada)
Sumitomo Bank of Canada
Swiss Bank Corporation (Canada)
Tokai Bank of Canada
Union Bank of Switzerland (Canada)
United Overseas Bank (Canada)
U.S. Bank (Canada)

Source: CBTA, *Bank Facts 1994*, pp. 21–3.

* There are 9 domestic bank CBA members and 58 foreign bank members; 43 foreign members are listed.

Appendix Eighteen

PEOPLE FOR AFFORDABLE TELEPHONE
SERVICE/COALITION POUR UN SERVICE
TÉLÉPHONIQUE AFFORDABLE: MEMBER
ORGANIZATIONS AS OF MARCH 1995

Association coopérative d'économie
  familiale de l'Estrie (ACEF)
ACEF de Granby
ACEF de Québec
ACEF de la Rive-Sud de Montréal
ACEF Centre de Montréal
ACEF de l' Outaouis
ACEF de Sud-Quest de Montréal
ACEF de Thedford-Mines
Action Centre for Social Justice
Alberta Council on Aging
ALERT PEI
Association des Consommateurs
  de Québec
Canadian Federation of Students
Canadian Grey Panthers
Canadian Pensioners Concerned
Canadian Pensioners Concerned
  (Alberta)
Canadian School Library Association
Canadian Teachers' Federation
Centrale de l'Enseignement
  du Québec
Centre de Pastorale en Mileu
  Ouvrier
Coalition for Public Information
  (CPI)
Comité des travailleurs et des
  travailleuses de l'Estrie
Confédération des Syndicates
  Nationaux
Consumers' Association of Canada
  (CAC)
CAC Alberta
CAC British Columbia
CAC Manitoba

CAC Newfoundland
CAC Nova Scotia
CAC North West Territories
CAC Quebec
CAC Saskatchewan
Council of Canadians
CYBER Québec Fédération des ACEF
Fédération des associations de fami-
  lies monoparentales du Québec
Féderation de l'Age d'Or du Québec
Féderation des Femmes du Québec
Féderation Nationale des Associa-
  tions de Consommateurs
  du Québec
Friends of Canadian Broadcasting
Front d'action populaire en
  réaménagement urbain
Income Security Action Committee
L'R des centres de femmes
  du Québec
Manitoba Society of Seniors
Mouvement d'education populaire
  et d'action communautraie
  du Québec
National Action Committee
  on the Status of Women (NAC)
National Anti-Poverty Organization
  (NAPO)
National Pensioners and Senior
  Citizen Federation
One Voice: The Canadian Seniors'
  Network
Ontario Coalition Against Poverty
Ontario Coalition of Senior
  Citizens' Organizations
People on Welfare for Equal Rights

Public Advisory Council on the
Information Highway
Regroupement des Comités
logements et des associations
de locations
Rural Dignity of Canada
Senior Link (Toronto)
Service budgétaire populaire
de l'Estrie

Service d'aide au Consommateur
de Shawinigan
Table d'action contre l'appauvrisse-
ment de l'Estrie
The Royal Canadian Legion
Vie ouvriére
Yukon Council on Aging

Source: PIAC Hotwire, issue 1, Mar. 1995,
p. 1.

Appendix Nineteen

COMPETITIVE TELECOMMUNICATIONS ASSOCIATION (CTA): MEMBERSHIP AS OF NOVEMBER 1996

ACC Long Distance Ltd.
AT&T Canada LDS
CamNet Communications Inc.
City Dial Network Services Ltd.
Consolidated Technologies Inc.
Fonorola Inc.
Funday Telecom
Hospitality Information Services Inc.
MetroNet Calgary
Northquest Telecom Inc.
Rogers Network Services
Sprint Canada Inc. (Call-Net Communications)
Smart Talk Network (STN)
TelRoute Communications Inc.
Unitel Communications Inc.
WesTel Telecommunications Ltd.
Whistler Telephone Co. Ltd.

Source: CTA Petition to Governor-in-Council, Government of Canada, "A Petition to Rescind Portions of Telecom Decision CRTC 94–19 and Telecom Decision CRTC 94–24," 23 Nov. 1994.

Appendix Twenty

ACTION CANADA NETWORK STEERING
COMMITTEE AS OF 1994

COMMUNITY ORGANIZATIONS

Canadian Federation of Students
Council of Canadians
Ecumenical Coalition for Economic
  Justice
National Action Committee
  on the Status of Women
National Anti-Poverty Organization
National Farmers Union
National Federation of Nurses
  Unions
National Pensioners and Senior
  Citizens Federation
Oxfam/CUSO/InterPares
Eastern Regional Coalition
  Representative
Western Regional Coalition
  Representative
Solidarité populaire Québec

LABOUR ORGANIZATIONS

Canadian Auto Workers
Canadian Labour Congress
Canadian Union of Postal Workers
Canadian Union of Public Employees
Confédération des syndicates
  nationaux
National Union of Public
  and General Employees
Public Service Alliance of Canada
United Steel Workers of America

REGIONAL COALITIONS

Action Canada Network Alberta
Action Canada Network BC

Action Canada Network PEI
Action Nova Scotia
Alternative North
Choices: a coalition for Social
  Justice
Fredericton and Area Coalition
  for Social Justice
Newfoundland Coalition for Equality
Ontario Coalition for Social Justice
Saskatchewan Coalition for Social
  Justice
Solidarité populaire Québec

PARTICIPATING NATIONAL
ORGANIZATIONS

Alliance of Canadian Cinema,
  Television and Radio Artists
Assembly of First Nations
Canadian Auto Workers
Canadian Centre for Policy
  Alternatives
Canadian Environmental Law
  Association
Canadian Federation of Students
Canadian Labour Congress
Canadian Peace Alliance
Canadian Teacher's Federation
Canadian Union of Public Employees
Common Frontiers
Communication Energy and
  Paperworkers Union
Confédération des syndicates
  nationaux
Confederation of Canadian Unions
Congress of Union Retirees
  of Canada
Council of Canadians

CUSO
Ecumenical Coalition for Economic Justice
Graphic Communications International Union
International Ladies Garment Workers Union
InterPares
Latin American Working Group
National Action Committee on the Status of Women
National Anti-Poverty Organization
National Farmers Union
National Federation of Nurses' Unions
National Pensioners and Senior Citizens Federation

National Union of Public and General Employees
Oxfam Canada
Public Service Alliance of Canada
Rural Dignity of Canada
Ten Days for World Development
Transportation-Communications International Union
United Church, Division of Mission in Canada
United Food and Commercial Workers
United Steelworkers of America
United Steelworkers of America, District 6
Voice of Women

Source: Action Canada Network, 1 Oct. 1994.

Appendix Twenty-one

ITAC/SAGIT TASK FORCE NAMES

INTERNATIONAL TRADE ADVISORY COMMITTEE (ITAC)

ITAC Task Force on International Business Development
ITAC Task Force on Trade and the Environment
ITAC Task Force on Trade Policy

SECTORAL ADVISORY GROUP ON INTERNATIONAL TRADE (SAGIT)

Advanced Manufacturing Technologies SAGIT
Agriculture, Food and Beverage SAGIT
Apparel and Footwear SAGIT
Business, Professional and Educational Services SAGIT
Cultural Industries SAGIT
Energy, Chemicals and Plastics SAGIT
Environmental Products and Services SAGIT
Fish and Sea Products SAGIT
Forest Products SAGIT
Household Products SAGIT
Information Technologies and Telecommunications SAGIT
Medical and Health Care Products and Services SAGIT
Mining, Metals and Minerals SAGIT
Textiles, Fur and Leather SAGIT
Transportation Equipment SAGIT
Transportation Services SAGIT

Source: Department of International Trade, 1995.

Appendix Twenty-two

INFORMATION TECHNOLOGIES
AND TELECOMMUNICATIONS
MEMBERSHIP LIST AS OF 1996
AND FORMER MEMBERS

CHAIRPERSON

Ms Janice Moyer, President, Information Technology Association of Canada

CURRENT MEMBERS

BCI, VP Peter Burn

Bell Canada, VP Bernard Courtois

Consortium UBI, Pres. Mme Sylvie Lalande

DEVELCON, VP George Best

Digital Equipment of Canada Ltd.,CEO Graeme Woodley

Digital Renaissance Inc., Pres. Keith Kocho

EyeTel Communications Inc., CEO Robert Calis

Global Business Alliance Inc., Pres. Peter Sandiford

Geomatics Industry Association of Canada, Pres. Ed. Kennedy

C.G. James and Associates, Pres. James Grant

MITEL Corporation, CEO Dr John Millard

NewEast Wireless Technologies Inc., Pres. Derrick Rowe

SHL Systemhouse International, CEO Jean-Pierre Soublier

Sierra Systems Inc., VP David Hughes

SR Telecom Inc., CEO Ronald Couchman

Telecommunications Workers Union, Pres. Rodney Hiebert

Teleglobe Inc., Executive VP Guthrie Stewart

Telesat Canada, CEO Larry Boisvert

TXN Soulition Integrators (subsidiary of Tandem Computers), Pres. Janice Moyer

FORMER MEMBERS

Alberta Research Council, Senior Scientist

Alcan Aluminum Ltd., Director of Information Services

Bell Canada, VP Law and Regulatory Matters

Canadian Cable Television Association, Pres.

Communications, Computer Equipment and Service SAGIT, former Chair

Communications, Energy and Paperworkers Union of Canada, Executive VP

Corporate Electronic Banking Services, Executive VP

De Loitte and Touche, Consulting Associate

DMR & Associates Ltd., Pres.

Dollard-des-Ormeaux, PQ

Financial Times of Canada, Publisher

Fonorola Inc., Pres. and CEO

Funday Cable Ltd., Pres.

Gandalf Technologies, Chair

Global Business Alliance Inc. Pres.

IBM Canada Ltd., VP, Law and Corporate Relations

International Brotherhood
of Electrical Workers,
International Representative
ISM-Information Systems Management Corp., VP Finance
C.G. James and Associates, Pres.
Leblanc and Royle Telecom Inc.,
Chair.
NewTel Enterprises Ltd., CEO
Novatron Information Corp., Pres.
Ogivar Technologies Inc., Pres.
Seabright Corp. Ltd., Pres. and CEO

Stentor Telecom Policy Inc., Pres.
and CEO
Telecommunications Workers Union,
Pres.
Teleglobe Inc. Executive VP
Telesat Canada, CEO
TGS Systems, Director, Finance
and Adm.
Zanthe Systems Division, Pres.

Source: Department of International Trade,
1996.

Appendix Twenty-three

THE ALLIANCE OF CANADIAN
COMMUNICATIONS UNIONS AS OF 1994

Atlantic Communications and Technical Workers' Union
Communications Energy and Paperworkers Union
Canadian Association of Communications and Allied Workers
International Brotherhood of Electrical Workers
National Federation of Communications Workers
Telecommunications Employees Association of Manitoba
Telecommunications Workers Union

Source: National Conference of Communications Unions, 1994; Atlantic Communications & Technical Workers' Union, Five Statements of Unity, July 1992.

Appendix Twenty-four

AN ALLIANCE
COMMUNICATIONS AND ELECTRICAL
WORKERS OF CANADA COMMUNICATIONS
WORKERS OF AMERICA SINDICATO
DE TELEFONISTAS DE LA REPUBLICA
MEXICANA

1 The three unions recognize the full autonomy and national rights of their members and organizations. This Alliance is meant to strengthen the respetive organizations and help establish a precedent for international solidarity with full national autonomy.

2 The Communications, Electrical and Publishing Industries are increasingly dominated by Trans National Corporations (TNCs) based in North America, Europe and Asia. This Alliance shall unite our organizations, when necessary, in order to strengthen the abilities of workers in Canada, the United States and Mexico to organized and bargain collectively.

3 In order to strengthen the ability of our members and our respective unions to both organize and collectively bargain, when necessary and possible, we will support joint mobilization in the countries affected to support those efforts.

4 Expanding national support with other labor organizations including our national labor centers, the Canadian Labour Congress and the AFL-CIO, will often be an important part of our efforts. Therefore, the CWC agrees to coordinate the efforts of both unions with the CLC, the CWA agrees to coordinate the effrots of both unions with the AFL-CIO, and the STRM agrees to coordinate the efforts of other trade union centers in Mexico.

5 Expanding political support becomes increasingly significant as the North American and world markets become increasingly integrated for both trade and envestment. Therefore, the CWC agrees to coordinate the political efforts of the three unions in Canada, the CWA will coordinate such efforts in the Unites States, while STRM will coordinate such efforts in Mexico.

6 CWC, CWA and STRM agree to work together to defend union and workers' rights in North America, including Mexico, and throughout the world. We will work to expand coordinated collective bargaining and organizing through other alliances and the appropriate international bodies.

7 CWC, CWA and STRM agree to maintain a working relationship for the permanent exchange of trade union information, knowledge and experience in reference to the respective areas in which their work.

Morton Bahr, President
Communications Workers
of America

Fred Pomery, President
Communications and Electrical
Workers of Canada

Francisco Hermandez Juarez
Sindicato de Telefonistas de la Republica Mexicana

Appendix Twenty-five

POSTAL, TELEGRAPH AND TELEPHONE INTERNATIONAL (PTTI): NORTH AMERICAN AND CARIBBEAN MEMBERS AS OF 1996

NORTH AMERICAN AFFILIATES

American Postal Workers Union
(APWU)

Baja California and Sonoira
Telephone Workers Union

Communication Electrical & Paper
Workers Union (CEP)

Canadian Union of Postal Workers
(CUPW)

Communications Workers
of America (CWA)

National Association of Letter
Carriers (NALC)

National Mail Handlers' Union
(NMHU)

Sindicato de Telefonistas de la
Republica Mexicana (STRM)

Telecommunications Workers Union
(TWU), British Columbia

Union of Postal Communications
Employees (UPCE)

CARIBBEAN AFFILIATES

Antigua Workers Union

Bahamas Communications and
Public Officers Union

Bahamas Public Services Association

Bahamas Communications and
Public Managers Union

Barbados National Union of Public
Workers

Belize Telecommunications Workers
Union

Public Service Union of Belize

Bermuda Public Service Association

Electricity Workers Union
of Curacao

Sindikato di Trahadonan den
Telekominikashon, Neth. Antilles

Waterfront and Allied Workers
Union of Dominica

Grenada Technical and Allied
Workers Union

Guyana Clerical and Commercial
Workers Union

Guyana Postal and Telecommunications Workers Union

Montserrat Allied Workers Union

St. Lucia Civil Service Association

Commercial, Technical & Allied
Workers Union of St. Vincent

Public Services Union of St. Vincent
and the Grenadines

Suriname Telephone and Telegraph
Workers' Union

Suriname Postal Workers' Union

Communications Workers' Union
of Trinidad and Tobago

Postal Workers' Union of Trinidad
and Tobago

Source: Postal, Telegraph and Telephone
International (PTTI), 1996.

Notes

INTRODUCTION

1 Over the past twenty-five years, in particular, a number of critical researchers have been documenting the continental regime change, the changes in the regime, and its impact on Canadian society (see Bleyer, 1992; Clement, 1977, 1994; Drache and Cameron, 1985; Jenson, 1989; Mahon, 1991; Richardson, 1992, among others.

CHAPTER ONE

1 According to Babe, high press rates and other abuses that can be traced to the vertical integration of carriage and content ended in 1910, when the Board of Railway Commissioners ruled that CP Telegraph's rates were discriminatory (1990: 58). It did not take very long after that for the telegraph company to withdraw from the news-gathering business.

2 As Babe (1990: 72) notes, the National Bell obtained 48.6 per cent of the controlling shares shortly after Bell Canada received its charter from the federal government in 1880. The irony of the situation is that, despite the First National Policy's focus on creating a national economy, National Bell's major interest and controlling shares in the Bell Telephone Company of Canada were met by a wall of silence and not even debated in the Senate or the House of Commons.

3 C.F. Sise, the executive director, along with his son C.F. Sise, Jr., "ruled Bell Canada and its manufacturing subsidiary ... from 1880 to 1944" (Surtees, 1994: 9).

4 Evidence presented at the select committee by F. Dagger reveals that AT&T held stock valued at $1,928,900 out of Bell's capital of $5,000,000. According to Dagger, telephone subscribers paid $3.60 per phone over and above the 8 per cent dividend that was expected by shareholders (Select Committee, 1905, vol. 1, app. 1, p. 9).

5 In an ironic twist Bell Nexia recently (2000) signed a contract with the Alberta government to provide high-speed data services by installing a fibre-optics network to various rural communities in the province. BCTelus (formerly Alberta Government Telephone), however, lost the bid.

6 Before the formation of TCTS Canadians relied on transmission through the U.S. for most of their trans-Canada telephone routings. When TCTS began constructing its link in 1932, it leased 25 per cent of its original 7,015 kilometre system from Canadian Pacific (Surtees, 1994:31).

CHAPTER TWO

1 The new industrialization, or second industrial revolution, as it is also known, is associated with technological advances in electric power and the internal combustion engine, among others.

2 The normal Fordist regime of regulation, developed by Aglietta (1979), Boyer (1990), Lipietz (1987), and others, provides an analysis of the European post-war reconstruction period. In most Western European countries Fordist institutions and national-wage setting was established by the central state through bilateral or trilateral measures, creating a Fordist compromise among labour, capital, and the state, whereas Fordist regulation that developed in both Canada and the United States was centred in the private sector, formulated through adversarial bargaining between labour and business (Drache, 1991: 251). In the Canadian and American Fordist compromise the state provides only supporting labour policies.

3 Winseck makes the argument that the IBEW was replaced, in part, with a company union because of its chauvinistic attitude towards the female members (1993: 89).

4 Glenday (1994: 27) incorrectly calls this competitive unionism. Social or co-operative unionism denotes a broader strategy that includes socio-economic reforms that are pursued by initiatives on both an economic and a political front.

CHAPTER THREE

1 William Coleman (1986: 137–8) identified 482 Canadian national associations representing capitalist interests as of 1980. These national associations are political agents for the various sectors of the Canadian

economy, from agriculture to communications, finance, manufacturing, resources, services, and trade, among others. By the 1990s capital political agents had proliferated to such an extent that the Canadian government passed the Lobbyist Registration Act (LRA) (1992) to keep track of these organizations. *The Federal Lobbyist 1995* breaks down these organizations into Tier I lobbyists, of which there are 800, and Tier II lobbyists, with approximately 2,400 registered lobbyists.

2 It is worth pointing out that the Conference Board of Canada is the subsidiary of a research organization whose headquarters is in the United States.

3 The Granscian term "organic intellectuals" is used to describe those academies and experts who, working with private and public research institutes, played a role in the production of hegemonic consent for the neo-liberal policy shift.

CHAPTER FOUR

1 Schiller's research reveals that the large telecommunications-user organizations and lobbyists kept reinventing themselves over and over again. Competitive Alternatives, which was formed in 1981, was an alliance of some of the same equipment manufacturers, service suppliers, and previous mentioned business users. Added to these were Independent Data Communications Manufacturers Association, ITT, General Telephone, Satellite Business Systems (the IBM consortium), and the Comsat, Aetna Life, Casualty consortium (1982: 90).

2 It is interesting to note that as a result of this consent decree the U.S. Department of Justice forced Western Electric, AT&T's telecommunications manufacturer, to divest itself of its Canadian subsidiary, Northern Electric. Total separation did not occur until 1970 (DOC, Minister of Communications, 1983: 8).

3 Although the Communication Workers of America helped to draft the bill, they did not specifically endorse it (Derthick and Quirk, 1985: 185).

4 In the Clyne Report sovereignty is defined as the ability of Canadians both in government and in the private sector to exercise control over the directions of economic, social, cultural, and political change (1979: 1).

5 PSTN system is a public network where data, voice, and non-telecom services are carried over a network using telephone signalling, local loops (local transmission), switching, and control apparatuses to interconnect local and long-distance circuits.

6 The CICA et al. included the Association of Competitive Telecommunication Suppliers, the Canadian Association of Data and Professional Service Organizations, the Canadian Bankers Association, the Canadian Business Equipment Manufacturers' Association, the Canadian Daily

Newspaper Publication Association, the Canadian Radio Common Carriers Association, Data Crown Inc., and the Telephone Answering Association of Canada. As telecommunications policy liberalization intensified in Canada, user organizations kept reinventing themselves in their various lobbying attempts.

CHAPTER FIVE

1 Although CNCP was not successful in this bid to enter and compete in the long-distance telephone service market, it was successful in its second attempt in 1992. After a series of name changes from Unitel to AT&T Canada, AT&T Canada (CNCP Communications with its United States partner) as of 1996 has obtained 9 per cent of the long-distance market (AT&T Canada submission to Telecom Public Notice CRTC 96–23, Mar. 1997, p. 27, Fig. 2).

2 NAPO's group members include women's groups, native women's councils, community-service councils, anti-poverty groups, churches, universities, food banks, co-operative-housing foundations, public-housing tenant councils, community health centres, welfare-rights coalitions, action centres for social justice, the Vanier Institute of the Family, and community legal clinics, among others (NAPO Group Members, 1997).

3 The CRTC approval of rate-of-return regulation for Bell and BC Telephone ranged from 10 to 14 per cent (Statistics Canada, *Telephone Statistics*, Cat. 56–303, 1980n–95).

4 PME stands for *petites moins enterprises*, in English the small-business community.

5 In just ten years from the time this document was released, telecommunications foreign ownership increased twice, first to 33 per cent ownership (chap. 4) and then to 46.7 per cent in 1997 (WTO, 1997).

6 As of 1994 there were over 200 domestic and foreign (read United States) VANs and resellers operating in every province in Canada except Alberta (Industry Canada, 1996: 54–6).

7 By-pass refers to the routing of telephone calls through networks other than a public switched telephone network to take advantage of cheaper rates. Many of the large Canadian telecommunications customers continually threatened to route their domestic and overseas long-distance calls through alternate carriers or company networks in the United States.

8 The acronym ITAC may be confused with the federal government committee the International Trade Advisory Committee, which is also referred to as ITAC (discussed in chap. 8). To eliminate confusion, ITAC will refer to the Information Technology Association and the government committee will be the International Trade Advisory Committee.

9 It will be recalled that the CCC raised $2.5 million for the two-year fight to ensure that long-distance telephone competition occurred. Similarly, the CBTA had access to approximately $1.5 to $2 million dollars per year in revenues (CBTA, *Annual Report*, 1990).

10 The telecom-user organizations refer to this as "regulatory balkaniza-tion" and a "patch-work of policies" (CBTA Telecom Alert, issue 2, p. 2; ITAC, 1989).

11 The Manitoba Government Telecom corporation was privatized by the Conservative government in December 1996 despite strong resistance from the opposition parties and numerous grassroots organizations formed to resist the privatization of MTS (Black and Mallea, 1987).

CHAPTER SIX

1 Unitel wanted permission to interconnect its system to Bell, BC Telephone, Island Tel, MT&T, NB Telephone, and Newfoundland Tel. BCRL wanted permission to interconnect its system to BC Telephone and Unitel.

2 It is important none the less to point out that Stentor does not always speak with one voice. Contrary views are held, for example, by AGT, as evidenced in the local measured service hearing (Rideout, 1996). It would appear that within Stentor new or contradictory positions are advanced by the organization with a more frontier attitude, AGT. These increasingly irreconcilable differences, in all likelihood, led to the break-up of Sentor.

3 These ideas are also put forward in *Talk Is Cheap: The Promise of Regulatory Reform in North American Telecommunications* by Robert Crandall and Leonard Waverman, in the 1995 Brookings Institution publication.

4 The OECD survey examines the prices for all types of services, including the fixed charges for business telephone lines, local charges, long-distance service, international long-distance service, and mobile commu-nications. Prices for each area are then indexed against the OECD aver-age. Using this extensive comparison, rather than the selective once used by the CCC, Canadian business subscribers are seen to have paid less for their basket of services than did subscribers in many other countries (1990: 51).

5 Bell proposed two phases of local rate increases. The first, an interim increase of $1.40 a month for residential subscribers, and $3.25 a month for business subscribers was slated to take effect 1 April 1993. The second increase was scheduled for 1 September 1993, with residen-tial subscribers paying an increase of $7.65 a month and business customers an increase of $15.35 a month (Bell, 1993: 2–3).

6 Although Bell's pro-competition position appears to contradict its previous position on liberalization and competition completely, it will be
recalled that in the first long-distance competition hearing and its
Macdonald Commission submission, Bell actually favoured competition
as long as regulatory constraints and obligations were loosened and it
could rebalance rates to reflect true costs.

7 Harry Trebling, in "Adapting Regulation to Tight Oligopoly," argues
that the telecommunications policy changes have in fact not achieved a
competitive marketplace but a tight oligopoly with industry-specific
barriers to entry, common corporate control of a differentiated market,
and the continued dominance of the incumbent firms (1995: 2).

8 Although the new telecom entrants captured approximately 30 per cent
of the market, the incumbent telephone companies' revenues had
increased to $9 billion by 1997 (Industry Canada, 1999: 20).

9 The Telecommunications Ombudsman provides a consumer-complaints
mechanism for the long-distance competitors and resellers such as ACC,
ACN, AT&T, City Dial, Consolidated Technologies, Rogers, Sprint,
and Westel.

10 The previous year MetroNet bought a national fibre infrastructure, indirectly, from partner Rogers Telecom Inc. These two arrangements permit
MetroNet to offer local and long-distance services, including voice, data,
and the Internet, to businesses in the largest Canadian cities (Lewis,
1999: D3).

11 Recently the U.S Justice Department approved a $61 billion merger
between SBC Corp. (formally South Western Bell) and Ameritech.

12 It is important to note that Spicer's comments were at odds with the
commissioners on the panel who rendered the decision of 92–12.

13 Stentor claimed that for every million dollars of investment in the
industry, it generated approximately eighteen to twenty jobs (Jocelyne
Côté-O'Hara, president and CEO of Stentor Telecom Policy Inc.,
15 December 1995: 2).

14 Support for the targeted-subsidy proposal came from the Alberta Council on Aging, the British Columbia Old Age Pensioners' Organization,
and others referred to as BCOAPO et al., the Consumers' Association
of Canada (Manitoba) and the Manitoba Society of Seniors Inc. (CAC/
MSOS), the Canadian Cable Television Association, the Information
Technology Association of Canada (ITAC), ACC Tel Enterprises, Sprint
Canada and Unitel Communications (now AT&T Canada), referred
to as the competitors, and the Manitoba Keewatiniwo Okimakanak
(MKO) community.

15 "The Survey of Consumer Perceptions Surrounding Telephone Service"
was conducted for and funded by the Public Interest Advocacy Centre,
Communications, Energy and Paperworkers Union of Canada, the

Telecommunications Workers Union, BC Public Interest Advocacy Centre, the Consumers' Association of Canada (national), and the Fédération nationale des associations des consommateurs du Québec.

16 The federal government's Connecting Canadians agenda includes numerous programs and initiatives designed to move Canada towards an information society and a knowledge economy. Because the focus of this agenda is on wiring and digitizing the whole country, existing affordability problems have a greater potential to mushroom.

CHAPTER SEVEN

1 Conference participants included Robert Crandall of the Brookings Institute; Lawson Hunter of Stikeman, Elliot; Tim Brenman of the University of Maryland; Marcel Boyer of the University of Montreal; Ian Angus of Angus TeleManagement; William Stanbury of the University of British Columbia; Greg Sidal of the American Enterprise Institute; Hudson Janisch of the University of Toronto; Bohdan Romoniuk from Alberta Government Telephone; and members of the telecommunications industry (Competition Bureau, *CompAct*, issue 2, Apr.-June 1996).

2 Although Treasury Board explains that the new department, Industry Canada, was a merger of the former Industry, Science and Technology, Consumer and Corporate Affairs, and the telecommunications policy and programs of the Department of Communications, former DOC employees referred to the reorganization as a takeover.

3 Initially not all members of the Canadian Manufacturers' Association supported free trade with the United States. Some, such as the beer industry and food manufacturers, among others, supported national policies to protect their markets. As free trade with the United States became a reality and their concerns were met, the CMA endorsed the trade-liberalization initiative.

4 Members of the Canadian Alliance for Trade and Job Opportunities included the BCNI, the Canadian Federation of Independent Business, the Canadian Manufacturers' Association, the Chamber of Commerce, and the national Consumers' Association of Canada (Doern and Tomlin, 1991: 216).

5 At the time of the free-trade election there were 13,000 members of the Council of Canadians. Concern for the neo-liberal policy shift has grown through the 1990s and the first year of the new century and is reflected in council membership, which as of 1997 was approximately 100,000 members (interview with D. Robinson, 1997).

6 Re-engineering is a term favoured by telecommunications corporations to describe their use of technology to radically restructure the workplace, including working conditions, type of work, hours, locations, and

the number of workers needed. A key component of re-engineering is getting the workers to accept these changes. What re-engineering really accomplishes is to eliminate approximately half of the work force through mechanization by shifting programs and service functions to the consumers (CWA, TIE-North America, 1994: 2).

7 Operator service profits are identified in a Bell-CEP joint study, but a previous agreement between the union and the company prevent its public release (G. Cwitco, 1999).

8 Negative billing occurred when the cable industry increased its cable rates by offering additional services. It was up to the cable customers to contact their cable service provider in order to refuse the new services and maintain their existing rates.

Bibliography

CANADIAN GOVERNMENT
DOCUMENTS AND PUBLICATIONS

1882. *Statutes of Canada.* Chap. 9, Sect. 4. May.

1969. The Government Organization Act. Ottawa: Queen's Printer.

1970. Railway Act, R.S.C. Sections 320 & 321. Ottawa.

1979. *Telecommunications and Canada: Consultative Committee on the Implications of Telecommunications for Canadian Sovereignty* (Clyne Report). Ottawa: Ministry of Supply and Services.

1982. *Royal Commission on the Economic Union and Development Prospects for Canada* (Macdonald Commission). Ottawa: Supply and Services

1984. *A Commission on Canada's Future: Challenges and Choices.* The Royal Commission on the Economic Union and Development Prospects for Canada (Macdonald Commission). Ottawa: Supply and Services.

1987. Teleglobe Canada Reorganization and Divesture Act. Ottawa: Queen's Printer.

1988. The Canada–U.S. Free Trade Agreement. Ottawa: Supply and Services.

1991. *Report to the People and Government of Canada.* Ottawa: Supply and Services.

1992. North American Free Trade Agreement. Between the Government of Canada, the Government of the United Mexican States, and the Government of the United States of America. Vol. 1, Chaps. 1–16, Ottawa: Supply and Services.

1993. Telecommunications Act. 40–41–42 Elizabeth II, Chap. 38. Assented 23 June; law 25 Oct.

Commons. 1902. *Debates*
- 1905. *Proceedings of the Select Committee on Telephone Systems.* Vol. 1 & 2, Ottawa: S.E. Dawson, Printer to the King's Most Excellent Majesty.
- 1992. Bill C-62, An Act respecting Telecommunications. Third session, Thirty-fourth Parliament 40–41, Elizabeth 11, 1991–92. First reading, Feb.
- 1993a. "Minutes of Proceedings and Evidence of the Sub-Committee on Bill C-62 of the Standing Committee on Communications and Culture." Issue no. 4, 29 Apr.
- 1993b. "Minutes of Proceedings and Evidence of the Sub-Committee on Bill C-62 of the Standing Committee on Communications and Culture." Issue no. 7, 6 May.
- 1998. Letter to Hon. Sheila Copps, P.C., M.P., Minister of Canadian Heritage, House of Commons, Ottawa, from thirty-seven M.P.S. 20 Apr.
Communications. 1969–1979. *Annual Report.*
- 1969. *Government Reorganization 1968: Final Report of Task Force 4.* 30 Jan.
- 1983. *Culture and Communications: Key Elements of Canada's Economic Future.* Brief submitted to the Honourable Francis Fox, Minister of Communications, to the Royal Commission on the Economic Union and Development Prospects for Canada. Montreal, Que. 3 Nov.
- 1985. DOC May 1985. Leaked Cabinet document. May.
- 1987. *Communications for the Twenty-First Century.* Ottawa: Minister of Supply and Services Canada.
- 1992. "Telecommunications: New Legislation for Canada." Feb.
- 1993. "Speaking Notes for the Honourable Perrin Beatty." *Communications.* P.S., M.P., Minister of Communications. Second reading of Bill C-62.
Director of Investigation and Research, Competition Act. 1991. "Evidence of the Director for CRTC Telecom Public Notice 1990–73." based on "The Prospective Benefits from Competition in the Canadian Long Distance Market" by Robert W. Crandall. 1 Mar.
External Affairs. 1986. "Minister for International Trade Announces Make-Up of International Trade Advisory Committee." *Communiqué.* 9 Jan.
Federal Court. 1992. *Bell Canada vs Unitel.* File A-900-92, Appeal Division 996–6795. 22 July.
Foreign Affairs and International Trade. ITAC and SAGITS background.
- 1986. "Minister for International Trade Announces the Formation of the Sectoral Advisory Groups and International Trade (SAGIT)." *Communiqué.* 3 Feb.
- 1996. Information Technologies and Telecommunications Membership List, Former Members of SAGIT, Information Technologies and Telecommunications List.

Human Resources. 1995. "Human Resource Study of the Canadian Telecommunication Industry: Human Resources Issue Paper." Submitted by KPMG, Pacific Leadership Inc., Tech Team Management Inc., Abt Associates, Interim Report. 30 Aug.
– 1996. *Human Resources Study of the Canadian Telecommunications Industry.* Summary Report. Ottawa: Ministry of Supply and Services.
Industry Canada. 1996a. *Annual Report, Director of Investigation and Research Competition Act.* For the year ended 31 Mar. Ottawa: Ministry of Supply and Services.
– 1996b. *The Telecommunications Service Industry: Trend Analysis Canada – United States 1980–1995.* Ottawa: Spectrum, Information Technologies and Telecommunications Sector of Industry Canada. Jan.
– 1996c. *CompAct.* News from the Competition Bureau. Issue 2, Apr.-June.
– 1997. Employment figures Compiled by Industry Canada.
– 1998. Contributions Program 1999–2000 Awards.
– 1999. *The Canadian Telecommunications Service Industry: An Overview 1997–1998.* Ottawa.
Privy Council. 1995. "Order-in-Council, P.C. 1995–2196." 19 Dec.
Senate. 1902. *Debates.*
– 1992. *Proceedings of the Standing Senate Committee on Transport and Communications.* Issue No. 16, 17, 25 May; No. 26 May; No. 18, 27 May.
Statistics Canada. 1947, 1950, 1955, 1960, 1965, 1970, 1975, 1980–95. *Telephone Statistics.* Cat. 56–303.
– 1960, 1970, 1980. *Household Facilities and Equipment.* Cat. 64–202.
– 1962, 1972, 1982, 1992. *Corporations and Labour Unions Return Act* (CALURA). Cat. 71–202.
– 1994. *Family Expenditure in Canada.* Cat. 62–555.
– 1995. *Historical Statistics of Canada.* Second ed. Ed. F.H. Leacy. Ottawa.
Supreme Court. 1989. *Alberta Government Telephone and the* CRTC *and* CNCP *Telecommunications and the Attorney General of Canada and the Attorneys General of Quebec, Nova Scotia, New Brunswick, Manitoba, British Columbia,Prince Edward Island, Saskatchewan, Alberta, and Newfoundland.* Ottawa: Supreme Court of Canada, File 19731. 14 Aug.
Telecommission. 1971. *Instant World: A Report on Telecommunications in Canada.* Ottawa: Information Canada.
Treasury Board. 1993. *Federal Regulatory Plan 1994.* Treasury Board Secretariat, Administrative Policy Branch, Regulatory Affairs. Ottawa: Ministry of Supply and Services.

OTHER DOCUMENTS AND PUBLICATIONS

Action Canada Network. 1994. National Co-Chairs, Program Committees, Participating Regional Coalitions, Participating National Organizations.

Aglietta, Michel. 1979. *A Theory of Capitalist Regulation: The U.S. Experience*. Trans. David Fernbach. London: Verso.

Ahrne, G., and W. Clement. 1994. "A New Regime? Class Representation within the Swedish State." In Clement and R. Mahon, eds., *Swedish Social Democracy: A Model in Transition*. Toronto: Canadian Scholars' Press Inc. 223–44.

Alford, Robert, and Roger Friedland. 1985. *Powers of Theory: Capitalism, the State, and Democracy*. New York: Cambridge University Press.

Althusser, L., and E. Balibar. 1968. *Reading Capital*. London: New Left Books.

Anonymous. 1997. Personal communication with former DOC employee, April.

Arthur Anderson & Co. 1979. *Cost of Government Regulation: Study for the Business Roudtable*. Chicago.

Anglo-Canadian Telephone Company. 1955–80. *Annual Report*.

Armstrong, Christopher, and H.V. Nelles. 1986. *Monopoly's Moment: The Organization and Regulation of Canadian Utilities, 1830–1930*. Toronto: University of Toronto Press.

AT&T Canada. 1997. "Comments of AT&T Canada Long Distance Service Company, Telecom Public Notice CRTC 96–26, Forbearance from Regulation of Toll Services Provided by Dominant Carriers." 26 March.

Atlantic Communication and Technical Workers' Union. 1992. "Five Statements of Unity as formulated by CEPU." July.

Aufderheide, Patricia. 1987. "Universal Service: Telephone Policy in the Public Interest." *Journal of Communication* 37(1) (Winter): 81–95.

Babe, Robert. 1988. "Control of Telephones: The Canadian Experience." *Canadian Journal of Communication* 13 (2): 16–29.

– 1990. *Telecommunications in Canada*. Toronto: University of Toronto Press.

– 1995. *Communication and the Transformation of Economics: Essays in Information, Public Policy, and Political Economy*. Boulder, Colo.: Westview Press.

Bell, Daniel. 1973. *The Coming of Post-Industrial Society*. New York: Basic Books.

Bell Canada. 1955, 1960 1965, 1970, 1975, 1980. *Annual Report*.

– 1983. "Bell Canada's Submission" to the Royal Commission on the Economic Union and Development Prospects for Canada, 14 Oct.

– 1993. "Canadian Radio-television and Telecommunications Commission, Bell Canada, General Increase in Rates, 1993, Part A, Request for Increase in Rates." 5 Feb.

– 1999. Newsroom Archives. *http:www.bell.ca/en/corp/aboutbell/newsroom/archives/nr99/orgchart.asp*.

Bernard, Elaine. 1982. *The Long-distance Feeling: A History of the Telecommunications Workers Union.* Vancouver: New Star Books.

Black, Errol, and Paula Mallea. 1997. "The Privatization of the Manitoba Telephone System." *Canadian Dimension* (Mar.- Apr.).

Bleyer, Peter. 1992. "Coalitions of Social Movements as Agents for Social Change: The Action Canada Network." In Carroll, William, ed., *Organizing Dissent: Contemporary Social Movements in Theory and Practice.* Toronto: Garamond Press. 102–17.

Bobbio, Norberto. 1979. "Gramsci and the Conception of Civil Society." In Mouffe, Chantal, ed., *Gramsci and Marxist Theory.* London: Routledge and Kegan Paul. 21–47.

Bolton, Brian. 1993. "Negotiating Structural and Technological Change in the Telecommunications Services in the United States." *Telecommunications Services: Negotiating Structural and Technological Change.* Geneva: ILO, 123–43.

Boyer, Robert. 1990. *The Regulation School: A Critical Introduction.* Trans. Craig Charney. New York: Columbia University Press.

Breton, Gilles, and Jane Jenson. 1991. "After Free Trade and Meech Lake: Quoi de neuf? *Studies in Political Economy* 34 (Spring): 199–218.

British Columbia Old Age Pensioners Organizations et al., Consumers' Association of Canada, Fédération Nationale des Associations de Consommateurs du Québec, Manitoba Society of Seniors, National Anti-Poverty Organization, One Voice – The Canadian Seniors' Network. 1996. "Perceptions of Telephone Service by Low Income Consumers." A Membership/Constituent Survey, Feb.

Brodie, Janine. 1989. "The Free Trade Election." *Studies in Political Economy* 28 (Spring): 175–82.

– 1990. *The Political Economy of Canadian Regionalism.* Toronto: Harcourt Brace Jovanovich.

Bruce, Robert, Jeffery Cunard, and Mark Director. 1986. *From Telecommunications to Electronic Services.* Washington, DC: Butterworths.

Buco-Glucksman, Christine. 1982. "Hegemony and Consent." In Sassoon, Anne Showstack, ed., *Approaches to Gramsci.* London: Writers and Readers Publishing Co-operative Society. 116–26.

Burnham, Peter. 1991. "Neo-Gramscian Hegemony and the International Order." *Capital and Class* 45 (Autumn): 73–93.

Business Council on National Issues. 1986, 1993, 1994/95. *Annual Report.*

Calabrese, Andrew, and Donald Jung. 1992. "Broadband Telecommunications in Rural America." *Telecommunications Policy* (April): 225–36.

Callinicos, Alex. 1985. "Anthony Giddens: A Contemporary Critique." *Theory and Society* 14 (2) (March): 133–66.

– 1987. *Making History: Agency, Structure and Change in Social Theory.* Cambridge, UK: Polity Press.

Call-Net Enterprises Inc. 1995–96. *Annual Report.*

Canadian Bankers Association. 1991. *American Long Distance Competition: The Power of Choice*. May.

Canadian Business Telecommunications Alliance. 1991. "Evidence of Canadian Business Telecommunications Alliance, CRTC Telecom Public Notice 1990–73." 1 Mar.

– 1991b. *Competition the Future for Canadian Telecommunications*. Toronto: CBTA.

– 1992. *Making the Connections: The CBTS's New Approach*. Jan.

– 1993. "The Telecommunications Act: A CBTA Reporting Service for Canadian Telecommunications Users." *Canadian Telecom Alert*. Toronto: CBTA.

– 1995. *1994–95 Annual Report*.

Canadian Centre for Policy Alternatives. 1992. *Which Way for the Americas? Analysis of NAFTA Proposals and the Impact on Canada*. In co-operation with Common Frontiers and Action Canada Network. Ottawa.

– 2001. *E-Commerce vs E-Commons*. Ed. M. Moll and L. Slade. Ottawa.

Canadian Chamber of Commerce. 1988. *Free Trade: The Canadian Chamber of Commerce Says Yes*. Ottawa.

Canadian Conference of Catholic Bishops. 1985. "The Potential of Public Enterprise." In Drache and Cameron, eds., *The Other Macdonald Report*.

Canadian Federation of Independent Business. 1991. "Ringing in the Savings: Small Business and Telecommunications Competition in Canada." Submission to the CRTC on the application of Unitel Communications Inc. and B.C. Rial Telecommunications/Lightel Inc. 27 Mar.

Canadian Radio-television and Telecommunications Commission (CRTC). 1977. Telecom Decision CRTC 77–16, "Challenge Communications Ltd. v. Bell Canada." Ottawa, 23 Dec.

– 1979. *Annual Report 1978–79*.

– 1979. Telecom Decision CRTC 79–11, "CNCP Telecommunications: Interconnection with Bell Canada." Ottawa, 17 May.

– 1981. Telecom Decision CRTC 81–24, "CNCP Telecommunications: Interconnection with the British Columbia Telephone Company." Ottawa, 24 Nov.

– 1982. Telecom Decision CRTC 82–14, "Attachment of Subscriber-Provided Terminal Equipment." Ottawa, 23 Nov.

– 1983. Telecom Decision CRTC 83–10, "Enhanced Services." Ottawa, 13 Oct.

– 1984a. Telecom Decision CRTC 84–10, "Radio Common Carrier Interconnection with Federally Regulated Telephone Companies." Ottawa, 22 Mar.

– 1984b. Telecom Decision CRTC 84–18, "Enhanced Services." Ottawa, 12 July.

– 1984c. Telecom Public Notice 1984–55. Hull, Que.

– 1985a. Telecom Decision CRTC 85–17, "Identification of Enhanced Services." Ottawa, 13 Aug.

– 1985b. Telecom Decision CRTC 85–19, "Interexchange Competition and Related Issues." Ottawa, 29 Aug.

- 1987a. Telecom Decision CRTC 87–1, "Interexchange Competition and Related Issues." Ottawa, 12 Feb.
- 1987b. Telecom Decision CRTC 87–13, "Cellular Radio – Adequacy of Structural Safeguards." Ottawa, 23 Sept.
- 1987c. Telecom Decision CRTC 87–1, "Resale to Provide Primary Exchange Voice Services." Ottawa, 12 Feb.
- 1988. *Competition in Public Long-Distance Telephone Service in Canada. Federal-Provincial-Territorial Task Force on Telecommunications Report* (Sherman Report). Ottawa: Minister of Supply and Services.
- 1992. Telecom Decision CRTC 92–12, "Competition in the Provision of Public Long Distance Voice Telephone Services and Related Resale and Sharing Issues." Ottawa, 12 June.
- 1992. "Consumer Friendly Competition: The Facts." Ottawa, 12 June.
- 1994. Telecom Decision CRTC 94–19, "Review of Regulatory Framework." Ottawa, 16 Sept.
- 1994. "Fact Sheet."
- 1996. Telecom Decision CRTC 96–10, "Local Service Pricing Options." Ottawa, 15 Nov.
- 1996. "Response to Interrogatory SRCI (CRTC)." 4Apr96–701.
- 1997a. "Local Competition." Telecom Decision CRTC 97–8.
- 1997b. "Service to High-Cost Areas." Telecom Public Notice CRTC 97–42.
- 1998. "Quarterly Monitoring Report." Telecom Order CRTC 97–1214. Stentor Resource Centre Inc., Apr.
- 1998. "Quarterly Monitoring Report." Telecom Order CRTC 97–1214. Stentor Resource Centre Inc., Aug.
- 1999a. "Telephone Service to High-Cost Serving Areas." Telecom Decision CRTC 99–16.
- 1999b. "Quarterly Monitoring Report." Telecom Order CRTC 97–1214. Submitted on behalf of BCT, Telus Communications Inc., Bell Canada, Island Telecom Inc., Maritime Tele & Tel Ltd., MTS Communications Inc., NBTel Inc., NewTel Communications Inc., Quebec-Telephone, Northwest Inc. Sept.
- 1999c. *Report on New Media.* Ottawa: CRTC.
- 2000a. CRTC *Action Plan 2000 2003.* http://www.crtc.gc.ca/eng/BACKGRND/pln20002.htm.
- 2000b. "Quarterly Monitoring Report." Telecom Order CRTC 97–1214. Submitted on behalf of Bell Canada, Island Telecom Inc., Maritime Tel & Tel Ltd., MTS Communications Inc., NBTel Inc., NewTel Communications Inc., Quebec-Telephone, Northwestel Inc., Telus Communications Inc. (B.C.), and Telus Communications Inc. Mar.

Canadians for Competitive Telecommunications. 1986. *The Crisis for Canadian Business: Telecommunications Rates and the Public Interest.* Feb.

Canadian Union of Public Employees. 1985. "The Potential of Public Enterprise." In Drache and Cameron, eds., *The Other Macdonald Report*. Toronto: James Lorimer & Company. 196–207.

Carleton Media and Communications Research Centre. 1990. "A Telecommunication Policy for All Canadians." Organized by V. Mosco of the School of Journalism and Communications and with funding from the Communication Workers of Canada.

Carroll, William. 1990. "Restructuring Capital, Reorganizing consent: Gramsci, Political Economy, and Canada." *Canadian Review of Sociology and Anthropology* 27 (3): 357–416.

Cavanagh, Richard. 1996. Personal communication with R. Cavanagh, Stentor Telecom Policy Inc., Director of Social Policy. 7 Feb.

Chaves, Peter. 1997. Personal communications with P. Chaves of AT&T Canada. 7 Apr.

Chomsky, Noam. 1993. "Notes on NAFTA 'The Masters of Mankind.'" *The Nation*, 29 Mar., 412–16.

Clark, Barry. 1991. *Political Economy: A Comparative* Approach. New York: Praeger.

Clement, Wallace. 1975. *The Canadian Corporate Elite*. Toronto: McCelland and Stewart.

– 1977. *Continental Corporate Power*. Toronto: McCelland and Stewart.

– 1983. *Class, Power and Property: Essays on Canadian Society*. Toronto: Methuen.

– 1988. *The Challenge of Class Analysis*. Ottawa: Carleton University Press.

– 1994. "Exploring the Limits of Social Democracy: Regime Change in Sweden." *Studies in Political Economy* 44 (Summer): 95–123.

Clement, Wallace, and Glen Williams, eds. 1989. *The New Canadian Political Economy*. Kingston & Montreal: McGill-Queen's University Press.

CNCP Telecommunications. 1983. "Telecommunications: A Core Component of Economic Growth." Submission by CNCP Telecommunications, with the assistance of Carl E. Beige, to the Royal Commission on the Economic Union and Development Prospects for Canada. Toronto, 31 Oct.

Cohen, Jeffrey. 1992. *The Politics of Telecommunications Regulation: The States and the Divesture of AT&T*. Armonk, NY: M.E. Sharpe.

Coleman, William 1986. "The Capitalist Class and the State: Changing Roles of Business Interest Associations." *Studies in Political Economy* 20 (Summer): 135–59.

Collins, Robert. 1977. *A Voice from Afar: The History of Telecommunications in Canada*. Toronto: McGraw-Hill Ryerson Ltd.

Colville, David. 1996. Personal communication with the vice-chairman of the Canadian Radio-television and Telecommunications Commission. 20 June.

– 2001. Personal communication, 5 Sept.

Communications, Electronic, Electrical, Technical and Salaried Workers of Canada (CWC). 1984. "Policy Document on Deregulation of the Telecommunications Industry: The Other Side of the Story." June.

Communications, Energy and Paperworkers Union of Canada (CEP). 1996. *Just the Beginning*. By James McCrostie. Ottawa.

Communications and Electrical Workers of Canada (CWC). 1991. "Final Argument of the Communications and Electrical Workers of Canada (CWC) before the Canadian Radio-television and Telecommunications Commission (CRTC) on Telecom Public Notice 1990–73." 29 July.

– 1992. "Comments on Bill C-62."

Communications Competition Coalition. 1991. "Evidence of M.N. Richardson, Communications Competition Coalition in the matter of: CRTC Telecom Public Notice 90–73." 1 Mar.

Communications Workers of America. 1994. *Preserving High-Wage Employment in Telecommunications*. CWA Public Policy Recommendations for a Competitive Regulatory Framework. May.

Communications Workers of America, Communications and Electrical Workers of Canada, Sindicato de Telefonistas de la Republica Mexicana. 1991. "An Alliance: Communications and Electrical Workers of Canada, Communications Workers of America, Sindicato de Telefonistas de la Republica Mexicana."

Competitive Telecommunications Association. 1994. Petition to the Governor in Council, Government of Canada, "A Petition to Rescind Portions of Telecom Decision CRTC 94–19 and Telecom Decision CRTC 94–24." 23 Nov.

Consumers' Association of Canada. 1986. "Emerging Telecommunications Issues: The CAC Perspective." Ottawa: Regulated Industries Program of the CAC, Feb.

– 1991. "Residential Telephony, Evidence of the Consumers' Association of Canada." Prepared by T.M. Denton Consultants Inc. Ottawa, Apr.

– 1996. "Tips '96."

Consumers' Federation of America. 1990. Speech at the conference , "A Telecommunication Policy for all Canadians."

Consumers' Union and Consumers' Federation of America. 2002. "The Telecommunications Act: Consumers Still Waiting for Better Phone and Cable Services on the Sixth Anniversary of National Law."

Coulter, B. Gerry. 1992. *The Emergence of a New Social Contract in Canadian Telecommunications Policy*. PhD, Carleton University, Ottawa.

Council of Canadians, Mel Clark. 1993a. "Restoring the Balance." Ottawa, Sept.

– 1993b. "Campaign for Canada: Stop Free Trade." *Canadian Perspectives*. Autumn.

Courchene, Thomas. 1985. "Privatization: Palliative or Panacea: An Interpretive Literature Survey." In W.T. Stanbury and T.E. Kierans, eds., *Papers on Privatization*. Montreal: Institute for Research on Public Policy. 1–36.

Courtois, Bernard. 1997. Personal communications with B. Courtois, Group Vice-President, Law & Regulatory Matters, Bell Canada. 20 June.

Cox, Robert. 1979. "Ideologies and the New International Economic Order: Reflections on Some Recent Literature." *International Organization* 33 (2): 257–302.

– 1987. *Production, Power, and World Order: Social Forces in the Making of History*. New York: Columbia University Press.

– 1992. "Global *Perestrokia.*" *Socialist Register 1992*. London: Merlin Press. 26–43

– 1993. "Structural Issues of Global Governance." In Stephen Gill, ed., *Gramsci: Historical Materialism and International Relations*. Cambridge, Mass.: Cambridge University Press. 259–89.

Crandall, Robert, and Leonard Waverman. 1995. *Talk Is Cheap: The Promise of Regulatory Reform in North American Telecommunications*. Washington, DC: The Brookings Institution.

Crandall, Robert, and Kenneth Flamm. 1991. *After the Breakup: U.S. Telecommunications in a More Competitive Era*. Washington, DC: The Brookings Institution.

Crow, Bob. 1996. Personal communications with B. Crow, director of research for Information Technology Association of Canada, 10 Mar.

Cruickshank Associates. 1989. *The Long Distance Competition Debate: Canadian Telecommunications in the Nineties*. Manotick, Ont.: D.J. Cruickshank Associates.

Dahl, Robert. 1956. *A Preface to Democratic Theory*. Chicago: Chicago University Press.

– 1982. *Dilemmas of Pluralist Democracy: Autonomy vs. Control*. New Haven, NJ: Yale University Press.

Derthick, Martha, and Paul Quirk. 1985. *The Politics of Deregulation*. Washington, DC: The Brookings Institute.

Dickinson, Paul, and George Sciadas. 1999. "Canadians Connected: Science and Technology Redesign Project." *Canadian Economic Observer*. Catalogue H-010-XPB, 3. 1–22.

Directory of Associations in Canada 1995–1996. 1996. Toronto: Micromedia Ltd.

Doern, G. Bruce, and Brian Tomlin. 1991. *Faith and Fear: The Free Trade Story*. Toronto: Stoddart.

Dordick, Nebert. 1990. "The Origins of Universal Service: History as a Determinant of Telecommunications." *Telecommunications Policy* 14 (3) (June): 223–31.

Drache, Daniel, and Duncan Cameron, eds. 1985. *The Other Macdonald Report*. Toronto: James Lorimer and Company.

Drache, Daniel, and Meric Gertler, eds. 1991. *The New Era of Global Competition: State Policy and Market Power*. Montreal and Kingston: McGill-Queen's University Press.

DuBoff, Richard. 1984. "The Rise of Communications Regulation: The Telegraph Industry, 1984–1880." *Journal of Communication* (Summer): 52–66.

Easton, David. 1965. *A Framework for Political Analysis.* Englewood Cliffs, NJ: Prentice Hall.

Economic Council of Canada. 1981. *Reforming Regulation.* Ottawa: Ministry of Supply and Services.

– 1982. *Lean Times.* Nineteenth Annual Review. Ottawa: Ministry of Supply and Services.

– 1992. *Pulling Together: Productivity, Innovation, and Trade.* Ottawa: Ministry of Supply and Services.

Ehrlich, Stanislaw. 1982. *Pluralism on and Off Course.* Oxford and New York: Pergamon Press.

Ekos Research Associates Inc. 1996. "Survey of Consumer Perceptions Surrounding Telephone Service." Submitted to Public Interest Advocacy Centre, Communications, Energy and Paperworkers Union of Canada, Telecommunications Workers Union, British Columbia Public Interest Advocacy Centre, Fédération des Associations de Consommateurs du Québec. 15 Feb.

– 2000. *Rethinking the Information Highway: Privacy, Access and Shifting Marketplace.* Ottawa: Ekos Research Associates.

Federal Communications Commission. 1960, 1980. *Statistics of Common Carriers,* Washington. DC.

– 1960, 1980. *Annual Report.* Washington, DC.

– 1981. "In the Matter of Policy and Rules Concerning Rules for Competitive Common Carrier Services." Docket No. 79–252, FCC.

The Federal Lobbyists 1995. 1995. Ed. John A. Chenier. Ottawa: ARC Publications.

Fédération Nationale des Associations de Consommateurs du Québec, the National Anti-Poverty Organization, and One Voice – The Canadian Seniors' Network. 1996. "In the Matter of Telecom Public Notices CRTC 95–49 and 95–56: Local Service Pricing Options." 19 Feb.

D.A. Ford and Associates. 1996. "Estimating the Costs of a Targeted Subsidy Plan for Canada." A report prepared for the Fédération Nationale des Associations de Consommateurs du Québec, the National Anti-Poverty Organization, and One Voice – The Canadian Seniors' Network. 30 Apr.

Fraser Institute. 1980. *Privatization Theory and Practice: Distributing Shares in Private and Public Enterprises.* By R. Ohashi, T. Roth, Z. Spinder, M. McMillen, and K. Norre. Vancouver.

Fritz, Rod. 1996. Personal communication with the president of Local 435 of the International Brotherhood of Electrical Workers representing workers at Manitoba Telephone System. 13 Sept.

Giddens, Anthony. 1984. *The Constitution of Society.* Berkeley and Los Angeles: University of California Press.

– 1987. *Social Theory and Modern Sociology.* Stanford, Calif.: Stanford University Press.

Gill, Stephen. 1993. "Epistemology, Ontology, and the 'Italian School.'" In Stephen Gill, ed., *Gramsci, Historical Materialism and International Relations.* Cambridge, Mass.: Cambridge University Press. 21–48.

Gill, Stephen, and David Law. 1988. *The Global Political Economy: Perspectives, Problems and Policies.* Baltimore, Md.: Johns Hopkins University Press.

– 1993. "Global Hegemony and the Structural Power of Capital." In Stephen Gill, ed., *Gramsci, Historical Materialism and International Relations.* Cambridge, Mass.: Cambridge University Press. 93–126.

Glenday, Dan. 1994. "On the Ropes: Can Unions in Canada Make a Comeback?" In Dan Glenday and Ann Duffy, eds., *Canadian Society: Understanding and Surviving the 1990s.* Toronto: McClelland & Stewart. 15–48.

– *Globe and Mail Report on Business.* 1985, 1995.

Globerman, Steven. 1986. "Economic Factors in Telecommunications Policy and Regulation." In W.T. Stanbury, ed., *Telecommunications Policy and Regulation: The Impact of Competition and Technological Change.* Montreal: Institute for Research on Public Policy.

Globerman, Steven. 1995. "Foreign Ownership in Telecommunications: A Policy Perspective." *Telecommunications Policy* 19 (1): 21–8.

Globerman, Steven, and Peter Booth. 1989. "The Canada-US Free Trade Agreement and the Telecommunications Industry." *Telecommunications Policy* (Dec): 319–28.

Globerman, Steven, with Diane Carter. 1988. *Telecommunications in Canada: An Analysis of Outlook and Trends.* Vancouver, BC: Fraser Institute.

Globerman, Steve, Hudson Janisch, and W.T. Stanbury. 1995. "Analysis of Telecommunication Decision 94–19, Review of Regulatory Framework." In S. Globerman, W.T. Stanbury, and T.A. Wilson, eds., *The Future of Telecommunications Policy in Canada.* Vancouver/Toronto: Bureau of Applied Research, Faculty of Commerce and Business Administration, University of British Columbia, and Institute of Policy Analysis, University of Toronto. 417–40.

Gramsci, Antonio. 1971. *Selections from the Prison Notebooks.* Ed. and trans. Q. Hoare and G. Smith. New York: International Publishers.

Granatstein, J.L. 1986. "Free Trade between Canada and the United States: The Issue That Will Not Go Away." In D. Stairs and G. Winham, eds., *The Politics of Canada's Economic Relationship with the United States.* Toronto: University of Toronto Press.

Grinspun, Ricardo, and Maxwell Cameron, eds., 1993. *The Political Economy of Free Trade.* Montreal and Kingston: McGill-Queen's University Press.

Haight, Timothy, and Laurie Weinstein. 1981. "Changing Ideology on Television by Changing Telecommunications Policy: Notes on a Contradictory

Situation." In Emile McAnany, Jorge Schnitman, and Noreene Janus, eds., *Communication and Social Structure*. New York: Praeger Publisher. 110–44.

Hamilton, Allen. 1996. Personal communication with A. Hamilton of Industry Canada, Telecommunication Policy Department. 19 Aug.

Hannigan, John. 1991. "Canadian Media Ownership and Control in an Age of Global Megamedia Empires." In B.D. Singer, ed., *Communications in Canadian Society*. Scarborough, Ont.: Nelson Canada. 257–75.

Hiebert, Rod. 1996. Personal communications with R. Hiebert, president of the Telecommunication Workers Union, 20 Feb.

Hiller, Harry. 1996. *Canadian Society: A Macro Analysis*. Third ed. Scarborough, Ont.: Prentice Hall Canada.

Hills, Jill. 1986. *Deregulating Telecoms: Competition and Control in the United States, Japan and Britain*. Westport, Conn.: Quorum Books.

– 1991. "Liberalization of Telecommunications in Britain and Its Impact on the Residential and Small Business Consumer." Evidence submitted for the TWU to the CRTC. Mar.

Horwitz, Robert B. 1989. *The Irony of Regulatory Reform: The Deregulation of American Telecommunications*. New York: Oxford University Press.

C.D. Howe Institute. 1976. *The Regulation of Private Economic Activity*. By Albert Breton. Toronto.

– 1980. *Conflict over Communications Policy: A Study of Federal-Provincial Relations and Public Policy*. Policy Commentary no. 1. R.B. Woodrow, K. Woodside, H. Wiseman, and J.B. Black. Toronto.

Howlett, Michael, and M. Ramesh. 1992. *The Political Economy of Canada: An Introduction*. Toronto: McClelland & Stewart.

Huber, Peter. 1993. "Telephones, Competition and the Candice-Coated Monopoly." *Regulation # 2*, CATO Review of Business and Government. 34–43.

Information Technology Association of Canada. "Backgrounder."

– 1989. *Telecommunications Regulation in Canada: A Patch-Work of Policies*. May.

– 1990. "An ITAC Statement on Competition Policy for the Telecommunications Industry." 17 Aug.

– 1993a. *Priorities for the Next Government of Canada*. Mississauga, Ont. June.

– 1993b. "New Telecommunications Legislation for Canada." A brief to the sub-committee on Bill C-62, Standing Committee on Communications and Culture, House of Commons. 3 May.

– 1994–95. *Annual Review, 1993–94, 1995*.

Innis, Harold. [1950], 1986. *Empire and Communications*. Intro. David Godfrey. Victoria/Toronto: Press Porcépic.

– 1956. "Decentralization and Democracy." *Essays in Canadian Economic History*. Ed. Mary Innis. Toronto: University of Toronto Press. 358–71.

Inter-Hemispheric Education Resource Center. 1992. *Cross-Border Links: A Directory of Organizations in Canada, Mexico, and the United States.* Ed. Ricardo Hernandez and Edith Sanchez. Albuquerque, NM.

International Labour Office (ILO). 1993. *Telecommunications Services: Negotiating Structural and Technological Change.* Contributors Brian Bolton, Edward Davis, Yann Landreau, Sean O'Ceallaigh, Norio Wada, Paul Willman. Geneva.

Intersect Alliance. 1998. "Representing the Consumer Interest into the Next Millennium: Recommendations for the Consumers' Association of Canada." Prepared for the Consumers' Association of Canada, 22 Oct.

Institute for Research on Public Policy. 1985. *Papers on Privatization.* Ed. T.E. Kierans and W.T. Stanbury. Montreal.

Janigan, Michael. 1997. Personal communication with M. Janigan, executive director of Public Interest Advocacy Centre. 11 Apr.

– 1989. "The Canada-US Free Trade Agreement: Impact on Telecommunications." *Telecommunications Policy* (June).

Janisch, H.N. 1986. "Winners and Losers: The Challenges Facing Telecommunications Regulation." In W.T. Stanbury, ed., *Telecommunications Policy and Regulation: The Impact of Competition and Technological Change.* Montreal: Institute for Research on Public Policy. 307–400.

Janisch, Hudson, and Schultz, Richard. 1985. "Teleglobe Canada: Cash Cow or White Elephant?" In T.E. Kierans and W.T. Stanbury, eds., *Papers on Privatization.* Montreal: Institute for Research on Public Policy. 185–242.

– 1989. *Exploiting the Information Revolution: Telecommunications Issues Positions for Canada.* Discussion paper commissioned by the Royal Bank of Canada. Oct.

Japan Economic Institute. 1992. "Participation of the Private Sector in U.S. Trade Policymaking: The North American Free Trade Agreement as Case Study." *JEI Report* No. 39A: 1–10.

Jenson, Jane. 1989. "Different but not Exceptional: Canada's Permeable Fordism." *Canadian Review of Sociology and Anthropology* 26 (1) (Feb.): 69–94.

Jessop, Bob. 1983. "Accumulation Strategies, State Forms, and Hegemonic Projects." *Kapitalistate* 10/11: 89–111.

Kerr, Jim. 1997. Personal communications with J. Kerr, Department of International Trade, chair of the SAGIT on Information Technologies & Telecommunications. 18 June.

Kierans, Eric. 1996. Personal communication with the former first minister of Communications. 30 Sept.

Kinkaid, James. 1996. Personal communication with national representative, Regulatory Affairs of the Communications, Energy and Paperworkers Union of Canada. 29 Aug.

Krasner, Stephen. 1984. "Approaches to the State: Alternative Conceptions and Historical Dynamics." *Comparative Politics* (Jan.): 223–46.

Laclau, Ernesto, and Chantel Mouffe. 1985. *Hegemony and Socialist Strategy: Towards a Radical Democratic Politics.* London: Verso.

Lang, Gary. 1990. Minister of Communications, Saskatchewan, speaker at the conference "A Telecommunication Policy for All Canadians."

Langille, David. 1987. "The Business Council on National Issues and the Canadian State." *Studies in Political Economy* 24 (Autumn): 41–85.

Lawson, Pilippa. 2001. *Eliminating Phoneless in Canada: Possible Approaches.* Public Interest Advocacy Centre.

Lipietz, Alain. 1987. *Mirages and Miracles: The Crisis of Global Fordism.* Trans. David Macey. London: Verso.

Lipsey, Richard. 1985. "Privatization: An Economist's Perspective." In T.E. Kierans and W.T. Stanbury, eds., *Papers on Privatization.* Montreal: Institute for Research on Public Policy. 37–46.

Lipsey, R.G., D. Schwanen, and R.J. Wonnacott. 1994. *The NAFTA: What's In, What's Out, What's Next.* Policy Study 21. Toronto: C.D. Howe Institute.

Mahon, Rianne. 1980. "Regulatory Agencies: Captive Agents or Hegemonic Apparatuses?" In J. Paul Grayson, ed., *Class, State, Ideology, and Change.* Toronto: Holt, Rinehart and Winston. 154–68.

– 1991. *Canadian Journal of Sociology.*

– 1991. "From 'Bringing' to 'Putting': The State in Late Twentieth Century Social Theory." *Canadian Journal of Sociology* (June): 119–44.

Manitoba Keewatinowi Okimakanak. 1996. "Telephone Affordability and Accessibility in the MKO First Nation Communities." Evidence of Marion Willis and Associates, William Kennedy Consultants Ltd., and Sean Keating. Feb.

Manitoba Telephone System. 1990. *People of Service: A Brief History of the Manitoba Telephone System.* Winnipeg, Man.: Corporate Communications, MTS

Mansell, Robin. 1993. *The New Telecommunications: A Political Economy of Network Evolution.* London: Sage.

Marchak, M. Patricia. 1991. *The Integrated Circus: The New Right and the Restructuring of Global Markets.* Montreal and Kingston: McGill-Queen's University Press.

McCall, Marnie. 1997. Personal communication with M. McCall, former director of Policy Research Association of the Consumers' of Canada. 7 Feb.

McEwan, Doug. 1997. Personal communication with D. McEwan, Telecommunications Policy, Industry Canada. 18 June.

McFarlane, Bruce. 1992. "Anthropologists and Sociologists, and Their Contributions to Policy in Canada." In W. Carroll, L. Christiansen-Ruffman, R. Currie, and D. Harrison, eds., *Fragile Truths: Twenty-Five Years of*

Sociology and Anthropology in Canada. Ottawa: Carleton University Press. 281–94.

McKendry, David. 1997. Personal communication with D. McKendry, formerly of Consumers' Association of Canada and director of the Regulated Industry Program. 17 Mar., 19 June.

– 2001. Personal communication, 5 Sept.

McPhail, Thomas and Brenda. 1990. *Communication: The Canadian Experience.* Toronto: Copp Clark Pitman Ltd.

Merrian, Charles. 1945. *Systematic Politics*, Chicago: University of Chicago Press.

Merton, Robert. 1949, 1968. *Social Theory and Social Structure.* New York: Free Press.

Miliband, Ralph. 1969, 1973. *The State in Capitalist Society.* London: Quartet.

Mintzberg, Henry. 1996. "Managing Government Governing Management." *Harvard Business Review* (May-June): 75–83.

Moll, Marita, and Leslie Regan Shade, ed. 2001. *E-commerce vs e-commons: Communications in the Public Interest.* Ottawa: Canadian Centre for Policy Alternatives.

Moody, Kim. 1998. "New Militancy and Leverage in Telecommunications: A Spirit of Resistance Grows." *http://twu@mail-list.com.* 23 Nov.

Moody, Kim, and Mary McGinn. 1992. *Unions and Free Trade: Solidarity vs Competition.* Detroit, Mich.: A Labour Notes Book. 64–9.

Mosco, Vincent. 1982. *Pushbutton Fantasies: Critical Perspectives on Videotext and Information Technology.* Norwood, NJ: Ablex Publishing.

– 1989. *The Pay-Per Society: Computers and Communication in the Information Age.* Toronto: Garamond Press.

– 1990a. "Towards a Transnational World Information Order: The Canada-U.S. Free Trade Agreement." *Canadian Journal of Communication* 15 (2) (May): 46–63.

– 1990b. *Transforming Telecommunications in Canada.* Ottawa: Centre for Policy Alternatives.

– 1991. "CRTC Telecom Public Notice 1991–73, Unitel Communications Inc. and B.C. Rail Telecommunications/Lightel Inc.: Applications to Provide Public Long-Distance Voice Service and Related Resale and Sharing Issues." Evidence of Professor Vincent Mosco on behalf of the Communications and Electrical Workers of Canada. Mar.

– 1993. "Transforming Telecommunications." In Janet Wasko, Vincent Mosco, and Manjunath Pendakur, eds., *Illuminating the Blindspot: Essays Honouring Dallas W. Smythe.* Norwood, NJ.: Ablex Publishing. 132–51.

– 1995. "Free Trade in Communication: Building a World Business Order." In Nordenstreng and Schiller., eds., *Beyond National Sovereignty.*

– 1996. *The Political Economy of Communication: Rethinking and Renewal.* Thousand Oaks, Calif.: Sage Publications.

– 1997. "Marketable Commodity or Public Good: The Conflict between Domestic and Foreign Communications Policy." In Gene Swimmer, ed., *How Ottawa Spends, 1997–98*. Ottawa: Carleton University Press. 159–78.

Mosco, Vincent, and Vanda Rideout. 1997. "Media Policy in North America." In John Corner, Philip Schlesinger, and Roger Silverstone, eds., *International Media Research: A Critical Survey*. London: Routledge. 154–83.

Mosco, Vincent, and Elia Zureik. 1987. *Computers in the Workplace*. Report to the Federal Department of Labour, Canada.

Nader, Ralph. 1992. Memo from Ralph Nader, Cable/TV Information Networks. 1 May.

National Anti-Poverty Organization. 1997. Mission Statement and Group Members.

National Conference of Communications Unions (Canada). 1994. "The Alliance." Participants List.

National Museum of Science and Technology. 1990. "Introduction of Communications Satellites in Canada." Interview with Eric Kierans, former first minister of Communications, conducted by Doris Jelly and Roy Dahoo. 28 Mar.

Noam, Eli. 1987. "The Public Telecommunications Network: A Concept in Transition." *Journal of Communication* 37 (1) (Winter): 30–48.

– 1992. *Telecommunications in Europe*. New York: Oxford University Press.

Nordenstreng, Kaarle, and Herbert Schiller, eds. 1993. *Beyond National Sovereignty: International Communication in the 1990s*. Second ed. Norwood, NJ: Ablex.

Oettinger, Anthony. 1988. *The Formula Is Everything: Costing and Pricing in Telecommunication Industry*. Cambridge, Mass.: Harvard University Program on Information Resources Policy.

Ogle, E.B. 1979. *Long Distance Please: The Story of the TransCanada Telephone System*. Toronto: Collins Publishers.

Olsen, Dennis. 1980. *The State Elite*. Toronto: McCelland and Stewart.

Ontario Public Service Union. 1985. "Jobs and the New Technology." In Drache and Cameron, eds., *The Other Macdonald Report*. 136–45.

Organization for Economic Co-Operation and Development. 1990. *Performance Indicators for Public Telecommunications Operators*. Paris.

Panitch, Leo. 1994. "Globalisation and the State." *Socialist Register 1994*. London: Merlin Press. 60–93.

Panitch, Leo, and Donald Swartz. 1985. *From Consent to Coercion: The Assault on Trade Union Freedoms*. Toronto: Garamond Press.

People for Affordable Telephone Service (PATS). 1995. "In the Matter of Telecom Decision 95-21 and in the Matter of Order in Council PC 1995-2196, Petition to Governor in Council Government of Canada."

– 1996. "Fair Share for All Customers." 15 Feb.

Pertschuck, Michail. 1982. *Revolt against Regulation: The Rise and Pause of the Consumer Movement*. Berkeley, Calif.: University of California.

Phillips, Almarin. 1991. "Changing Markets and Institutional Inertia: A Review of US Telecommunications Policy." *Telecommunications Policy* 15 (1) (Feb.): 49–61.

Pike, Robert, and Vincent Mosco. 1986. "Canadian Consumers and Telephone Pricing: From Luxury to Necessity and Back Again?" *Telecommunications Policy* 10 (1) (March): 17–32.

Pool, Ithiel de Sola. 1983. *Technologies of Freedom*. Cambridge, Mass.: Belknap Press.

Postal, Telegraph and Telephone International. 1996. "Postal, Telegraph and Telephone International (PTTI) Background."

Poulantzas, Nicos. 1973. "On Social Classes." *New Left Review* 78: 25–54.

– 1974. *Classes in Contemporary Capitalism*. London: New Left Books.

– 1978. *State, Power, Socialism*. London: New Left Books.

Public Interest Advocacy Centre. 1994. "The Regulation of Telecommunications in Canada: A Consumer Perspective." By Philippa Lawson, with funding from Industry Canada. 22 Feb.

– 1995. *Hotwire*. Mar.

– 1996. "Consumer Coalitions: Three Case Studies." Prepared by M. Janigan, P. Lawson, M. Vallée, and A. Reddick.

– 1998. "Still a Long Distance To Go: Residential Consumers and the Transition to Competition in the Long Distance Market." Written by Angie Barrados. Feb.

Raboy, Marc. 1990. *Missed Opportunities: The Story of Canada's Broadcasting Policy*. Montreal and Kingston: McGill-Queen's University Press.

Reddick, Andrew. 1993. "Banking, Communications and Information Technology." MA, Ottawa: Carleton University.

– 1996. "Property Rights and Communication." *Alternate Routes* 13: 23–65.

– 2001. Personal communication, 10 Dec.

Reddick, Andrew, with C. Boucher and M. Groseilliers. 2000. *The Dual Digital Divide: The Information Highway in Canada*. Ottawa: Public Interest Advocacy Centre.

Rens, Jean-Guy. 2001. *The Invisible Empire: A History of the Telecommunications Industry in Canada, 1846–1956*. Trans. Käthe Roth. Montreal and Kingston: McGill-Queen's University Press.

Richardson, Jack. 1992. "Free Trade: Why Did It Happen?" *Canadian Review of Sociology and Anthropology* 29 (3): 307–28.

Richardson, Monty. 1996. Personal communications with M. Richardson, executive director of the Communications Competition Coalition. 11 May.

Rideout, Vanda. 1991. "Canadian Telecommunication Public Policy: A Study in Political Economy." MA, Ottawa: Carleton University.

– 1993. "Telecommunication Policy for Whom? An Analysis of Recent CRTC Decisions." *Alternate Routes* 10: 27–56.

– 1996. "Summary of Issues at the Local Service Pricing Options CRTC Hearing." Final Report, 17 June.

Rideout, Vanda, and Vincent Mosco. 1997. "Communication Policy in the United States." In Bailie and Winseck, eds., *Democratizing Communication? Comparative Perspectives on Information and Power.* Cresskill, NJ: Hampton Press, Inc. 81–104.

Robertson, Dr Gerald. 1997. Personal communication with Dr Gerald Robertson, co-ordinator, Regulatory Economics, Economics and International Affairs Branch, Competition Bureau, Industry Canada. 7 Feb.

Robinson, Dave. 1995, 1997. Personal communications with D. Robinson, director of Research of the Council of Canadians. 7 Feb. 1995, 23 June 1997.

Robinson, Ian. 1993. *North American Trade: As If Democracy Mattered.* Ottawa: Canadian Centre for Policy Alternatives.

Roman, Andrew. 1990. "The Telecommunication Policy Void in Canada." *Canadian Journal of Communication* 15 (2): 96–110.

Schiller, Dan. 1982. *Telematics and Government.* Norwood, NJ: Ablex Publishing.

– 1985. "The Emerging Global Grid: Planning for What?" *Media, Culture and Society* 7: 105–25.

Schiller, Dan, and Rosa Fregoso. 1995. "A Private View of the Digital World." In Nordenstreng and Schiller, eds., *Beyond National Sovereignty.* 210–34.

Schiller, Herbert. 1989. *Culture, Inc.: The Corporate Takeover of Public Expression.* New York, Oxford: Oxford University Press.

Schmitter, Philippe. 1979. "Still the Century of Corporatism?" In P. Schmitter and G. Lehmbruch, eds., *Trends Toward Corporatist Intermediation.* London: Sage Publications. 7–52.

Schultz, Richard. 1995. "Old Wine in a New Bottle: The Politics of Cross-Subsidies in Canadian Telecommunications." In S. Globerman, W.T. Stanbury, and T.A. Wilson. eds., *The Future of Telecommunications Policy.* Vancouver and Toronto: University of British Columbia and University of Toronto. 271–88.

– 1988. "Forward to the Past: The Canadian Approach to Telecommunications Regulatory Reform." Prepared for University of Vermont–McGill University Conference "Managing Global Telecommunications Politics: North American Perspectives." Burlington, Vt. Working Paper 1988–42.

Schultz, Richard, and Alan Alexandroff. 1985. *Economic Regulation and the Federal System.* Toronto: University of Toronto Press.

Schultz, Richard, and Hudson Janisch. 1993. *Freedom to Compete: Reforming the Canadian Telecommunications Regulatory System.* Discussion paper commissioned by Bell Canada. Ottawa: Bell Canada. Mar.

Science Council of Canada. 1982. *Planning Now for an Information Society: Tomorrow Is Too Late.* Ottawa: Ministry of Supply and Services.

Shefrin, Ivan. 1993. "The North American Free Trade Agreement: Telecommunications in Perspective." *Telecommunications Policy* 17 (1) (Jan./Feb.): 14–26.

Shniad, Sid. 1996. Personal communication with S. Shniad, director of Research, Telecommunications Workers Union. 14 Feb.

Showstack Sassoon, Anne, ed. 1982. *Approaches to Gramsci*. London: Writers and Readers Publishing Co-operative Society.

– 1987. *Gramsci's Politics*. Second ed. London: Hutchinson.

Simeon, Richard. 1987. "Inside the Macdonald Commission." *Studies in Political Economy* 22 (Spring): 167–79.

Simpson, Carl. 1996. Personal communication with the vice-president of the Atlantic Communications and Technical Workers Union. 12 Sept.

Sinclair, Allison. 1996. Personal communication with director of Research, the Business Council on National Issues. 20 Feb.

Skocpol, Theda. 1985. "Bringing the State Back In: Strategies of Analysis in Current Research." In Peter Evans, D. Rueschemeyer, and Theda Skocpol, eds., *Bringing the State Back In*. New York: Cambridge University Press. 3–43.

Smythe, Dallas. 1981. *Dependency Road: Communications, Capitalism, Consciousness, and Canada*. Norwood, NJ: Ablex Publishing.

Social Planning Council of Metropolitan Toronto. 1985. "Economic Decline in Canada." In Drache and Cameron, eds. *The Other Macdonald Report*. 3–22.

Sprint Annual Report. 1995, 1996.

Stanbury, W.T. 1986. "Decision-Making in Telecommunications: The Interplay of Distributional and Efficiency Considerations." In W.T. Stanbury, ed., *Telecommunications Policy and Regulation: The Impact of Competition and Technology Change*. Montreal: Institute for Research on Public Policy. 481–516.

– 1993. *Business-Government Relations in Canada: Influencing Public Policy*. Second ed. Scarborough, Ont.: Nelson Canada.

– 1995. "Redeeming the Promise of Confluence: Analysis of Issues Facing the CRTC." In S. Globerman, W.T. Stanbury, and T.A. Wilson, eds., *The Future of Telecommunications Policy in Canada*. Vancouver and Toronto: University of British Columbia and University of Toronto. 441–88.

Stanbury, W.T., and Fred Thompson. 1982. *Regulatory Reform in Canada*. Montreal: Institute for Research on Public Policy

Stentor Telecom Policy Inc. 1995. Letter written by Ms. Jocelyne Côté-O'Hara, president and CEO of Stentor, to the Clerk of Privy Council. 15 Dec.

– 1996. "Submission of Stentor Resource Centre, Inc. Local Service Pricing Options, Telecom Public Notice CRTC 95–49." 19 Feb.

Stentor Telecom Resource Centre Inc. 1998. "1997 Annual Monitoring Report." Telecom Order CRTC 97–1214: Annual Monitoring Report. 21 Dec.

Surtees, Lawrence. 1994. *Wire Wars: The Canadian Fight for Competition in Telecommunications*. Scarborough, Ont.: Prentice Hall.

Taylor, Allan. 1989. "Canada's Future Telecom Policy." Speech prepared by chairman and CEO of the Royal Bank of Canada for delivery to the Canadian Club of Toronto. 16 Oct. *Transnational Data and Communications Report*, 18–20.

Telecommunications Workers Union. 1984. "Competition in Canada's Telephone Industry?" The Telecommunications Workers Union Position Paper on the Upcoming CRTC Hearings. Spring.

– 1990. "The Future of Canada's Telecommunications System: Canadians Have a Choice." A position paper by the Telecommunications Workers Union on the Rogers/Unitel application to the CRTC for permission to sell public long-distance telephone service. Winter.

– 1996. "Building Alliances – The Key to the Future of the Labour Movement." A presentation by Rod Hiebert, president of the Telecommunications Workers Union, to the Postal, Telegraph and Telephone International, XIV Inter-American Congress, San Jose, Costa Rica. 20–23 Aug.

Texier, Jacques. 1979. "Gramsci, Theoretician of the Superstructures." In C. Mouffe, ed., *Gramsci and Marxist Theory*. London: Routledge and Kegan Paul.

Thrift, Nigel J. 1985. "Bear and Mouse or Bear and Tree? Anthony Giddens's Reconstitution of Social Theory." *Sociology* 19 (4): 609–23.

Tinker, Toney, C. Leham, and M. Neimark. 1988. "Bookkeeping for Capitalism: The Mystery of Accounting for Unequal Exchange." In V. Mosco and J. Wasko eds., *The Political Economy of Information*. Madison, WI: University of Wisconsin Press. 188–216.

Toronto Star. 1997. "Voice mail lets the 'phoneless' keep in touch." 15 May, B6.

Toupin, Lynn. 1997. Personal communications with director of National Anti-Poverty Organization, 6 May.

Transnationals Information Exchange (TIE-North America). 1994. "US-Mexico-Canada Telecommunication Workers' Conference Report." Oaxtepec, Mexico. 10–13 Feb.

Trebling, Harry. 1995. "Adapting Regulation to Tight Oligopoly." Draft unpublished paper presented to the CRTC. Michigan State University.

Tritt, Robert. 1997. Personal communication with national director, International Affairs, Stentor Telecom Policy Inc. 23 June.

Truman, David B. 1951. *The Government Process*. New York: Alfred Knopf.

United Auto Workers (Canadian Auto Workers). 1985. "Can Canada Compete?" In Drache and Cameron, eds., *The Other Macdonald Report*. 23–34.

United Church of Canada. 1985. "Economic Development and Social Justice." In Drache and Cameron, eds., *The Other Macdonald Report*. 169–83.

Vallée, Marie. 1997. Personal communications with the director of Research for Fédération des associations de consommateurs de Québec. 10 Apr.

van der Pijl, Kees. 1984. *The Making of an Atlantic Ruling Class*. London: Verso.

van Koughnett, Greg. 1997. Personal communications with the vice-president for Stentor, Legal and Corporate Affairs. 9 Apr.

Warnock, John. 1988. *Free Trade and the New Right Agenda*. Vancouver: New Star Books.

Weinhaus, Carol, and Anthony Oettinger. 1988. *Behind the Telephone Debates*. Norwood, NJ: Albex Publishing.

Wendt, Alexander. 1987. "The Agent-Structure Problem in International Relations Theory." *International Organization* 41 (3) (Summer): 335–70.

Who's Who in Canadian Business. 1996. Toronto: Trans-Canada Press.

Williams, Raymond. 1977. *Marxism and Literature*. Oxford: Oxford University Press.

Marion Willis and Associates, William Kennedy Consultants Ltd., and Sean Keating. 1996. "Telephone Affordability and Accessibility in the MKO First Nation Communities." A report prepared for the Manitoba Keewatinowi Olimakanak Inc. (MKO) for CRTC Public Notice 95–49, Local Service Pricing Options. Feb.

Wilson, James. 1980. *The Politics of Regulation*. New York: Basic Books.

Wilson, Kevin. 1988. *Technologies of Control: The New Interactive Media for the Home*. Madison, Wisc.: University of Wisconsin Press.

– 1992. "Deregulating Telecommunications and the Problem of Natural Monopoly: A Critique of Economics in Telecommunications Policy." *Media, Culture and Society*. 343–68..

Winseck, Dwayne. 1993. "A Study of the (De)Regulatory Process in Canadian Telecommunication: Labour Struggles and the Public Interest." PhD, University of Oregon, Eugene.

– 1995. "A Social History of Canadian Telecommunications." *Canadian Journal of Communication* 20 (2): 1–20.

– 1998. *Reconvergence: A Political Economy of Telecommunications in Canada*. Cresskill, NJ: Hampton Press.

Winter, James, and Amir Hassanpour. 1994. "Building Babel." *Canadian Forum* (Jan./Feb.): 10–17.

Woodrow, R. Brian, Kenneth Woodside, Henry Wiseman, and John B. Black. 1980. *Conflict over Communication Policy: A Study of Federal-Provincial Relations and Public Policy*. C.D. Howe Institute.

Woodrow, R. Brian, and Ken Woodside. 1986. "Players, Stakes and Politics in the Future of Telecommunications Regulation in Canada." In W.T. Stanbury, ed., *Telecommunications Policy and Regulation: The Impact of Competition and Technological Change*. Montreal: Institute for Research on Public Policy. 101–249.

World Trade Organization. 1998a. "Audiovisual Services." S/c/w/40 98–24–37. June 1–19.

– 1998b. *Electronic Commerce and the Role of the WTO*. Geneva: WTO Publications.

Index